Outdoor Storage

Created and designed by the
editorial staff of ORTHO Books

Project Editor	Diane Snow
Writer	Margaret Lucke
Construction Consultant	Bob Beckstrom
Designer	Craig Bergquist
Illustrator	Rik Olson
Photographers	Laurie A. Black
	Kathryn Kleinman
Photographic Stylist	Sara Slavin

Ortho Books

Publisher
Robert L. Iacopi

Editorial Director
Min S. Yee

Managing Editors
Anne Coolman
Michael D. Smith

System Manager
Mark Zielinski

Senior Editor
Sally W. Smith

Editors
Jim Beley
Diane Snow
Deni Stein

System Assistant
William F. Yusavage

Production Manager
Laurie Sheldon

Photographers
Laurie A. Black
Michael D. McKinley

Photo Editors
Anne Dickson-Pederson
Pam Peirce

Production Editor
Alice E. Mace

Production Assistant
Darcie S. Furlan

National Sales Manager
Garry P. Wellman

Operations/Distribution
William T. Pletcher

Operations Assistant
Donna M. White

Administrative Assistant
Georgiann Wright

Address all inquiries to:
Ortho Books
Chevron Chemical Company
Consumer Products Division
575 Market Street
San Francisco, CA 94105

Chevron Chemical Company
575 Market Street, San Francisco, CA 94105

Acknowledgments

Photography
Laurie A. Black
Pages 1, 4, 6, 7 (bottom), 8–9 (top),
  9 (bottom), 38, 52
Barbara J. Ferguson
Front cover, page 7 (top)
Kathryn Kleinman
Pages 24–37, 70–71

Photographic Locations
B. Gay Ballard
Walnut Creek, CA
Kathryn Hadley
San Francisco, CA
Susan W. Plimpton
North Kingstown, RI

Designers
Page 1: Bob Johnson,
  Woodside, CA
Pages 4, 6, 7(bottom): Donald G.
  Boos, Landscape Architect,
  Los Altos, CA
Pages 8–9: Casey A. Kawamoto,
  Landscape Architect,
  San Francisco, CA
Page 38: David Shuter,
  Joshua Tree, CA

Consultants
Arrow Group Industries
Pompton Plains, NJ
East-West Designs, Inc.
Madison, WI
Randall Fleming, Architect
Anthony/Fleming & Associates
Oakland, CA
T. Jeff Williams
Potter Valley, CA

Special Thanks
Mrs. T. Bacigalupi, Jr.
Julie S. Bartrum
Mr. and Mrs. Erich Berthold
Shelley and David Hamel
Gregory J. Hauca
Don Hunter
Ken Jung
Jerry and Laura Kahn
Jack Kay
Denise Kelly
Mr. and Mrs. Victor Morvay
Susan Nelson
Mr. and Mrs. Paul Weidemann
Mark and Debbie Wrighton

Backen, Arrigoni & Ross
San Francisco, CA
Bourriague Construction
Santa Cruz, CA
California Redwood Association
Mill Valley, CA
Callander Associates
Burlingame, CA
Callister, Gately & Bischoff
Tiburon, CA
Daily Construction Service
San Francisco, CA
Emery Rogers & Associates
San Francisco, CA
Grounds Maintenance
Overland Park, KS
Hardware Age
Los Angeles, CA
Harry C. Bond & Associates
Napa, CA
Home Center Institute
Indianapolis, IN
Home Center Magazine
Chicago, IL

James Ransohoff & Associates
Menlo Park, CA
Lehmann Landscaping Co., Inc.
San Mateo, CA
Nova Research Group
San Francisco, CA
Mathew W. Henning,
  Landscape Architect
Oakland, CA
Matsutani & Co., Inc.
Concord, CA
Richard Murray Associates
Carmel, CA
Robert M. Babcock & Associates
Lafayette, CA
Western Landscaping
San Mateo, CA
Winfield Smith & Associates
Glen Ellen, CA

Photographic Props
Brown Jordan Co.
Floorcraft
Fillamento
Forrest Jones, Inc.
Presidio Bicycle Shop
Dave Sullivan Sports

Editorial Assistant
Karin Shakery

Typesetting
CBM Type
Sunnyvale, CA

Color Separation
Color Tech
Redwood City, CA

Front Cover:
Emphasizing the country feel of a large semi-rural lot, this barn-styled shed is home to a riding mower, rakes, and other yard tools. Painted a deep green, the shed blends into the surrounding environment of lawn and trees. The window admits light, the vent above it provides needed air circulation, and the gambrel roof allows room for an overhead loft that holds items used less frequently.

Page 1:
Nestled in a leafy bower, this storage shed enhances the appearance and enjoyment of the swimming pool. It keeps the beach ball, swim rings, and raft ready for fun, while the convenient spot for storing skimmers and chemicals makes maintenance an easy chore. Extended with a fence and arbor, yet open to the view of trees, the structure defines the pool area comfortably, separating it from the forest beyond.

Back Cover:
Planning storage according to your own living patterns and spatial needs can produce attractive, effective solutions. One approach (top left) is to custom-build shelves or cabinetry, designed to suit your needs and dimensioned to fit your existing space. Another approach (top right) is to buy manufactured products. Small-scale components (bottom left) quickly help you make the most of storage areas. Large-scale storage structures, such as metal sheds (bottom right), help you enclose needed storage volume and space.

Outdoor Storage

Planning for Successful Storage

Planning will give you the most effective storage system for your particular household. Look at the examples showing individual approaches to storage space. Then follow a seven-step process to arrive at your own solution.

What to do with the lawnmower, the luggage, the bicycles, the barbecue grill, the canning kettle, the cushions to the patio chairs. We all face problems with storing the myriad of possessions that are crowding us out of—well, if not house and home, at least carport and garage.

Sometimes it seems as though someone forgot, when the house was built, that "home" means both outdoor and indoor activities—along with all the equipment they include. Things used inside spill outdoors because there isn't enough room for them in the house. And outdoor activities, frequently accompanied by heavy, bulky, or odd-sized equipment, require additional storage space. Hundreds of large and small items must be kept in some kind of order. Yet your home may have no outdoor storage area. Or its storage area may have no closets, cabinets, or shelves. Therefore, not only do you have to organize all these items, but you must also create places to put them. The purpose of this book is to help you find solutions to these problems quickly and effectively.

How to Use This Book
The book is structured to help you plan and create a storage system that suits your particular home and life-style. The *system* is simply the logic behind your storage decisions: You match items with storage space in a way that supports and simplifies the flow of your activities.

The first chapter deals with planning—the all-important foundation of any storage system.

A careful evaluation of storage priorities led to this solution for a family that entertains often: a hutch for an outdoor dining room. The base cabinets store utensils for *al fresco* meals, the blue-tiled counter is a handy serving surface, and the open shelves allow for displays that are both festive and functional. Appearance was as much a design issue as capacity was; the result is a practical storage space that harmonizes well with the deck and arbor.

Planning helps you avoid moving things from one corner to another without really solving your storage problems. The scope of your plan can be large or small, depending on your needs and your level of ambition.

The second chapter is a photographic survey of storage elements you can buy—hooks, hangers, shelves, bins, and other items for organizing your things quickly and easily. The third chapter illustrates a selection of projects to build if you prefer a more customized approach. If these smaller-scale components won't suffice, you may decide to add a storage structure to your property. The fourth chapter presents an overview of some of the prefabricated buildings and shed construction kits that are available, along with advice on assembling and outfitting them. The last chapter provides plans for sheds and other structures you can build from scratch.

Getting Started
Just tackling the issue of storage is the first step—and often the first hurdle as well. Yet dealing with storage problems is a particularly satisfying endeavor, because clearing out household bottlenecks frees you from a nagging backlog of unfinished business: "Someday I've got to clean out the garage. Someday I'm going to organize all this stuff so I can find things. Someday . . . ." When you decide that "someday" has arrived, and you begin finding solutions to these bothersome little problems, the result is often so rewarding that you wonder why you waited so long to deal with them.

Begin by looking at the photos and illustrations on the next few pages. They'll fire your imagination with ideas about what can be accomplished when you pay a little attention to outdoor storage. Then launch into planning your own storage system using the guidelines in the balance of this chapter.

Blending Storage In

A storage structure doesn't have to pop up in your yard like a mushroom, or stick out like a sore thumb. It can blend harmoniously and attractively with your outdoor space. The storage areas shown on these four pages don't call attention to themselves; in some cases they're hard to spot at all. Yet they're right there, convenient and ready to go to work when needed.

Sitting quietly in a service yard, this shed almost disappears behind the fence, despite the open gateway. From a distance, it seems to be a continuous part of the fence. Two factors in the design contribute to this effect. First, both structures are faced with the same material, and painted the same color. Second, the fence was built to the same height as the shed's eaves to create an unbroken visual line across the top.

Tucked underneath the raised deck, a nearly invisible storage compartment makes use of space that might otherwise be wasted. Its slatted sides don't totally shut out the elements, though the space can still be put to good storage uses. The enclosure might accommodate trash cans, firewood on a pallet, or other items that can stand a little exposure. Its location, just a few steps down from the back door, makes this storage nook especially handy.

Bumping out from both sides of the fence, this sizable shed shows that an ample and serviceable structure does not have to clash with the handsome style of a home. The shed's siding materials repeat those of the fence, while its proportions echo those of the garage. Both structures are further unified by the dark brown edging that caps all upper edges and links the parts together. Low bushes line the fence, while taller shrubbery in front of the shed reflects its increase in scale. On the far side of the fence, the shed doors open onto a patio.

Blending Storage In

Storage stays in plain view in this patio trellis. What looks like a gate in the middle is really the door to a small shed housing barbecue equipment and garden-party supplies. By thinking in broader terms when they planned the trellis screen, the homeowners gained extra storage space in addition to privacy, shade, and an attractive focal point for their outdoor living area. The shed is shorter than the lattice panels and its front isn't crisscrossed with lathing, yet it blends into the overall structure, thanks to its shared color, similar materials, and centered position.

The simple storage cabinet shown below makes good use of the fence (which conceals it from the neighbors' view). Built of the same material as the fence, and painted the same color, the cabinet does its job without being obtrusive—proving that a structure can blend in even if it isn't elegant or elaborate. The sloped roof sheds rain; each roof panel overlaps the front, making finger-lifts to swing the panels up on rear-mounted hinges. This gives access to the contents from the top as well as the front. A thick gravel bed and paving blocks inside keep the ground from getting muddy.

Shaping Your Space

Illustrated on these six pages are three typical outdoor storage spaces—a garage, a carport, and a prefabricated shed—each outfitted in three different ways to serve the needs of various households.

These three areas are ones which, in many homes, are often not put to use as effectively as they could be. While this is sometimes due to inadequate space, you might be overlooking potential storage areas. For example, there is often space overhead that is underutilized.

Once you zero in on the problems and give thought to the possible solutions, you will be amazed at how a huge jumble can be reorganized into a relatively small space. Not only will the floor suddenly be swept clear, you will be able to spend the afternoon digging in the garden instead of through a mountain of mess in order to find the spade.

The particular objects stored in these spaces are incidental. What is important is that individual storage requirements can be successfully met, even when working within the confines of one type of space, simply by developing a storage plan. Many of the storage components shown in these solutions can be found in later chapters of the book.

Garages: Reorganizing Existing Space

Making the best use of space you already have is the most direct and economical route to solving your storage problems. For many people, that space is a garage. Because it's an empty shell, things tend to collect there willy-nilly. Finding things, not to mention reaching them once you have found them, can be awkward and difficult. This tedious state of affairs can be resolved with a little attention to the nature of the problem, a bit of thoughtful planning, and some Saturday morning elbow grease.

Take a look at some of the ways to put a two-car garage to work more efficiently. The three households that use these garages have some common goals: They all want room inside the garage to park the car, and they all want to provide space for an activity—woodworking, auto maintenance, potting plants—as well as for storage. Because their activities and living patterns differ, they organize and equip their spaces in different ways. In each case, however, the result is a system that allows family members to locate, use, and put away their possessions with a minimum of fuss—and, as a result, to get more fun out of their activities.

Multiple-Purpose Storage

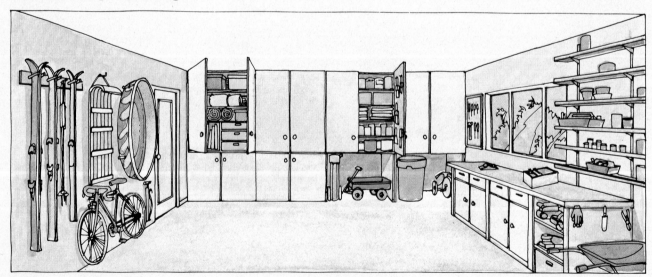

This garage is outfitted to reflect the multiple needs and interests typical of many households. One person enjoys carpentry, another gardening. Gear is stowed for all sorts of outings, such as weekend camping trips and ski vacations. A combination of closed cabinets keeps infrequently used items neat, easy to see, easy to reach, and dust free. Open shelves and pegboard hooks keep often-used objects handy yet organized. Hazardous items, such as insecticides and sharp tools, are stored high, out of the reach of small children. The shelves holding the garden chemicals and sprayer have lips around the edges to reduce the risk that a bottle will be accidentally knocked off. Bulky, wheeled items are stored against the walls, or in a special compartment beneath the cabinets. They don't block access to anything else, yet leave plenty of room for the car.

Combination Work and Storage Area

These family members share a strong interest in cars and engines, so their garage is more than just a parking spot; it has been transformed into a full-scale center for automotive and mechanical work. A bench across the back wall provides a work area, and space below it accommodates awkward items: tires, a dolly, even an engine block awaiting repair. Because there are few other cabinets or low shelves, plenty of floor space is available for the motor vehicles.

High shelves and wall surfaces provide places for the equipment used in this family's other favorite activity: get-away-from-it-all outings. Racks and hooks hold the fishing gear. Bulky but lightweight camping equipment has a home up near the cross-ties. Even the ties are put to good use, holding paddles, a canoe, and a camper shell for the pickup truck.

Storage for Outdoor Activities

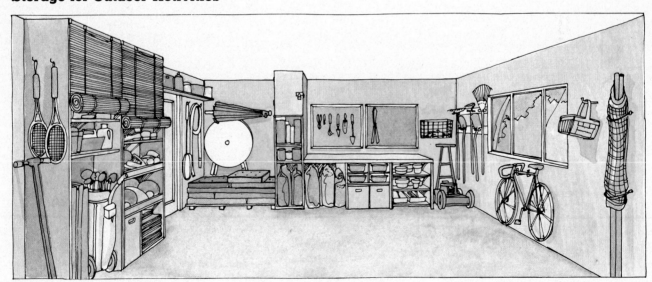

This family enjoys summer sports and outdoor entertaining. During the winter, sports equipment and patio paraphernalia find a dry home in the garage. The open-fronted cabinets hold playing balls, golf clubs and shoes, water toys, plus an assortment of household items. The bamboo shades can be lowered to present a neat appearance (see page 47). Swimming pool equipment and an umbrella table share the nook behind the door. While the heavy redwood patio chairs and lounge remain outdoors under plastic covers, their cushions are sheltered here during the off-season. Plywood resting on 4 by 4s keeps them away from dust and dampness on the floor.

The potting bench is the most active area in the garage. To its left, a shelf unit adds storage space, and a locked cabinet at the top stores garden poisons safely.

Shaping Your Space

Carports: Creating Enclosed Spaces

A carport is an open structure, and that fact affects its use for storage in two ways. One is functional, and one is aesthetic.

First, a carport does not enclose and secure things. Any item left there is exposed and vulnerable. To protect belongings in a carport, you need to add a lockable structure, such as a cupboard, cabinet, or closet, that will shelter your things from damaging weather and secure them from theft.

Second, any change made to a carport will be highly visible, so you must give extra consideration to style and design. The storage structure you install becomes a new element in the architectural detailing of your home. You want it to look like an integral part of the house and grounds, not like a tacked-on

afterthought. Using compatible colors, materials, and types of trim will help you achieve this. Pay attention also to the lines and proportions of your storage enclosure. If the project is a major one, the professional eye of an architect or contractor may be helpful in the planning stage.

The type of storage structure you build for your carport is best determined as part of your overall planning for a personal storage system. What kinds of things do you need to store? How often do you need them? A small overhead cabinet isn't a practical spot for a heavy item or something that's in daily use. Nor do you need to wall in the whole carport in order to hang up a few gardening tools. An enclosure that looks beautiful but doesn't help your household function more smoothly will be a source of frustration more than of pride.

These examples show three varied approaches to the same double carport. The additions that were made were carefully planned to meet two objectives: They fit into their surroundings in a pleasing way, and they create new storage space, tailored to each family's particular requirements.

In the carport below, the entire back wall is given over to a bank of shallow closets. This way, the addition of storage only minimally affects the amount of floor space. In the upper example on the facing page, cupboards are hung around the perimeter. They are securely fastened to both the posts and the ceiling members. On the facing page, below, is an example of why it is necessary to order your priorities. Here, it was more important to provide a workshop than to protect a second car from the elements.

A Wall of Storage

Adding a storage wall gives visual closure to this carport. The backyard gains privacy without sacrificing easy access to the rear of the house.

The storage wall consists of four modular cabinets. The lawnmower and tricycle sit on the floor. Eye-level

shelves and pegboard hangers make it easy to keep track of small items. The backs of the cabinet doors give lots of ready-access storage, and the narrower shelves inside allow them plenty of room to close. Objects are visible and reachable, but the family

still has the boon of capacious storage areas.

Each module has been assigned a category. One is for patio and pool equipment. A second is for garden supplies. The module closest to the house contains overflow from inside.

Overhead Storage Modules

A sleek and handsome set of cupboards maintains this carport's open feeling while adding a considerable amount of storage capacity (see page 48). The family's aim was to gain storage space without changing the original structure's clean lines and styling. In addition, all of the floor space is reserved for the family cars.

While very large, heavy objects can't be stored here, most of the supplies needed for auto care, gardening, and an assortment of leisure activities have found a home. The stepladder, which can be hung on the wall, provides access; while lockable cabinet doors discourage unauthorized borrowers.

Workshop and Storage Space

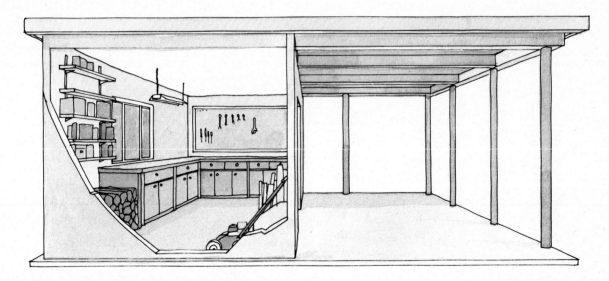

This household needs to keep only one car out of the weather, so half the carport has been enclosed. A new room—part workshop, part storeroom—provides functional space. An oversized sliding door allows activities to expand into the car space when needed (see page 92). Large materials and tools on casters move in and out easily. A U-shaped workbench occupies one end, and wooden storage shelves mounted on Z-brackets hang above a stack of firewood at the other end. Floor space is shared by a riding mower, a scrap lumber bin, and a heavy-duty power saw. Long-handled tools hang on one wall.

Outside, the enclosure is faced with the same material as the house and painted to match, so that it is compatible with the original structure.

Shaping Your Space

Sheds: Adding Another Structure

Many people put up prefabricated sheds in the hope that having more space will be the answer to their storage problem. But, without planning, a shed might simply give your problem extra room into which to expand. It's important to decide how you want the new structure to work for you before you set it up and move your belongings into it.

Storage space is not simply a place where things sit when they are not in use. Consider it in active terms—it's a place where items move in and out, a place that makes it easier (or harder) to accomplish the activities that you use the items for.

How objects flow in and out of their storage locations is one of the key issues you deal with in the course of developing your storage plan. As you figure out

how you want to set up storage areas, consider the age, size, and physical capacities of the people who use an item—for instance, how high can they reach? Also think about how often an object is used—the most frequently used belongings should be the most accessible. The seasonal use of many items is another factor in outdoor storage. Something used almost daily in summer might not even be looked at during the snowy months; it may need a new, out-of-the-way home during the winter.

In the three situations shown here, each household has chosen a standard 10-foot by 9-foot steel shed and set it up on a concrete slab. The three sheds function effectively yet quite differently, thanks to precise planning of what items are to be stored inside, how they should be stored inside, how they should be stored to facilitate the

flow in and out, and what storage accessories are therefore required.

Sometimes you can get good ideas from the color brochures and advertisements that shed manufacturers produce to promote their products. However, these companies can only guess at what you will be storing and how you want to organize. It is you who knows the specific needs. Put your ingenuity to work and you will be surprised at how easy it is to make every cubic foot work to your best possible advantage.

A metal shed is not easy to modify or customize, but because these families gave a lot of thought to the role the shed could play in their storage systems, their structures successfully meet their individual needs.

The rest of this chapter outlines a seven-step planning process that will guide you in developing your own storage system.

Providing Household Storage Outdoors

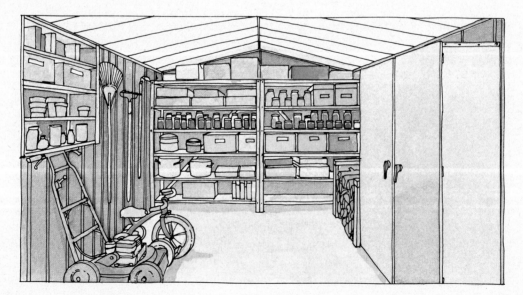

When a house has no carport or garage, a shed can provide a place to store a good deal of household overflow. If no family member is interested in building a structure, a prefabricated shed can be purchased and completely outfitted with storage components available at local stores. Across the back of this one, free-standing metal utility shelves house supplies for gardening and home maintenance as well as cartons of canning jars. A steel cabinet keeps out-of-season clothing from jamming the closets inside the house.

Firewood is stacked neatly in a corner, and the tricycle and lawnmower are wheeled against the walls. The floor space in the center is kept clear, so that household members can easily reach all the stored items.

Storing Space-Consuming Equipment

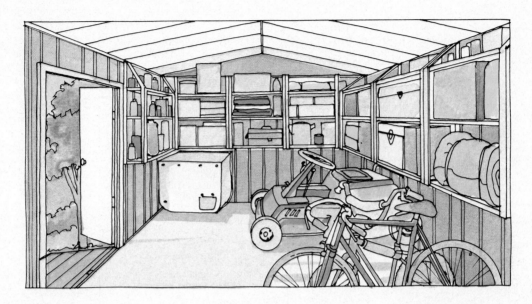

Big, awkward, wheeled items were taking up so much space in this family's garage that there was no room for the car. Now the riding mower, wheelbarrow, and kids' bikes are neatly stowed in a row on the floor of a separate shed. To make use of the space above, the walls are outfitted with accessory shelf units sold by the shed manufacturer. The shelves hold garden tools, chemicals and gear for bicycle trips.

Doubling Up on Storage Capacity

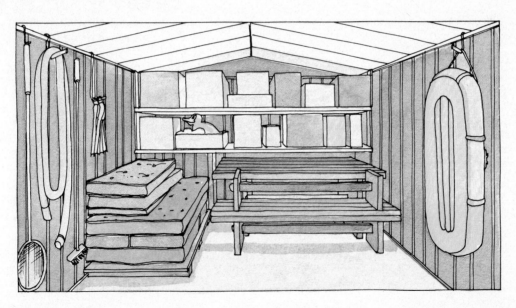

When nights turn frosty, this family packs its summer fun equipment in the shed, where it is protected until the following spring. Picnic tables and patio furniture are carefully stacked. The grill and the child's wagon are rolled out of the way. Swim toys and supplies for open-air parties are put on shelves. During winter months, the equipment is left undisturbed, so access is not an issue.

When it's time for beaches and barbecues again, the storage shed is emptied of its summer contents, and winter equipment—which had been shifted to the garage for easy access—moves back in. The shed is still a convenient spot to keep entertainment supplies and outdoor toys and to protect chair cushions when it begins to rain.

Seven Steps to Successful Storage

At first glance it may seem that storage is simply a matter of finding more space to store things. But unless there's an organizational system that ties all the elements together and makes sense in terms of all your activities over the course of a year, it won't be long before you're tripping over tangles of hose, searching in dusty corners for the screwdriver or the picnic cooler, and wondering how things got so messy again. The planning process outlined here is designed to help you create an outdoor storage system that is logical and workable for your entire household all year long.

1. Sort and Toss

Much of your storage problem might be solved just by eliminating the objects that are making no useful contribution to your household. Why put up shelves for outgrown ice skates, a broken lawn chair, or a tower of empty boxes?

If you're overwhelmed by a hodgepodge of garden tools, paint cans, bikes, snow tires, and skis, the feeling will dissolve as you sort through things one by one, deciding either to keep or to toss them. All you need is an ample supply of cartons and heavy-duty garbage bags.

To start, roll up your sleeves, dig in, and have fun with it. Sort and toss quickly; don't labor over the details. Chances are you'll find many things you haven't seen—or used—in years. You can dispose of these without a second thought.

When Something Has Value

People often find it difficult to get rid of old things because they attach value to them—monetary value, sentimental value, or potential future value. While this is valid, it doesn't take into account the value of your time, your space, and the comfortable flow of your life. Measured against these more immediate values, the worth of those old ice skates goes way down. It doesn't mean they're worthless; it means their value to *you* has diminished over time.

Monetary value. Items that have monetary value can be sold at a garage sale or an auction, or taken to an antique dealer or a second-hand store. Or you might give them away to relatives, friends, or charities.

Sentimental value. Items that have sentimental value are often the toughest to toss. When you sort through them, you'll be reminded of significant times in your life. This can be a forceful experience that makes you cling to the objects along with the memories. One way to continue enjoying the memories, yet reduce your storage load, is to keep a few things that characterize these special moments, and let the rest of the items go. That way, you can continue to enjoy important parts of your past, without letting them inhibit the flow of your current life.

Future value. A common storage mistake is to hang onto things because they "might come in handy someday." You may be saving supplies for a hobby, long put aside, on the theory that perhaps you'll take it up again. But even if you do take it up, chances are you'll want to buy new equipment anyway. If you can't bear to toss it, consider "loaning" the equipment to a friend or relative with the idea that you can borrow it when you need to use it.

"Someday" repair projects also invite a closer look. Realistically, when is fixing that broken chair going to be your top priority on a Saturday?

Clearing Things Up and Out

You may feel it is wasteful to throw things away, but the benefits of organization and extra space can easily exceed the replacement value of a few lumber scraps or bicycle tires. Use your sense of reality—and your sense of humor—to get over those moments of hesitation. The trepidation you feel when you toss things out or give them away will most likely be followed by liberation and relief.

Once you've sorted through all your belongings, dispose of everything you're not going to keep. Call the charity or disposal service; set the date for the garage sale. It's easy to put this off, leaving boxes and bags of discarded items hanging around your garage for another year, and getting in your way as you try to devise a workable storage system. If removal can't take place right away, put all the discards in one area where they won't block your path as you work out a new storage plan for those possessions that are truly useful or enjoyable.

Sort into Categories

When you're reacquainted with the things you want to keep and store, categorically sort them into piles. Typical categories are gardening items, auto equipment, home maintenance supplies, outdoor entertainment furniture and equipment, sporting goods and toys, workbench tools, and household overflow (seasonal clothing, holiday decorations, card tables, cots, and other things that don't fit into your closets and cupboards inside). Sorting gives you a clearer picture of which categories have the most items, which ones have only a few, which ones need special storage conditions, etc. Then let the piles sit for a while as you analyze the space you have—or may create—to store them in.

2. Sketch a Simple Site Plan

An outdoor storage system works best when it's a natural outgrowth of your family's living patterns. Think of outdoor storage as part of your master plan for the whole house, accommodating outdoor equipment and supplies as well as the overflow from your inside storage areas.

The basis of the plan is a simple aerial view sketch of your house and yard. This sketch helps you see where you engage in particular activities, which, in turn, helps you determine the ideal spots for storing

items related to those activities. It's not critical to have a scale drawing—you just want a sketch precise enough to show the physical relationships of each area to the next and the traffic paths that connect them. However, if you already have a site plan drawn to scale (you may have received one when you bought the house), or if you want to take the time to make one, by all means use it—you won't misjudge sizes and distances, and your planning will be more accurate.

On your sketch, include the basic floor plan of the house, showing windows and exterior

doors. Include the carport or garage and any other outbuildings, along with special features of your landscape: fence, deck, patio, driveway, walkways, play areas, sandbox, swing set, trees, vegetable, and flower beds.

This sketch, by the way, is also a very helpful tool for other home improvement projects. Because it gives you an overview of your entire site and the features and elements within it, the sketch can help you make new landscape and garden plans, remodeling plans, and much more. It is, therefore, worth doing a site plan just to have it on hand.

Sample Site Plan

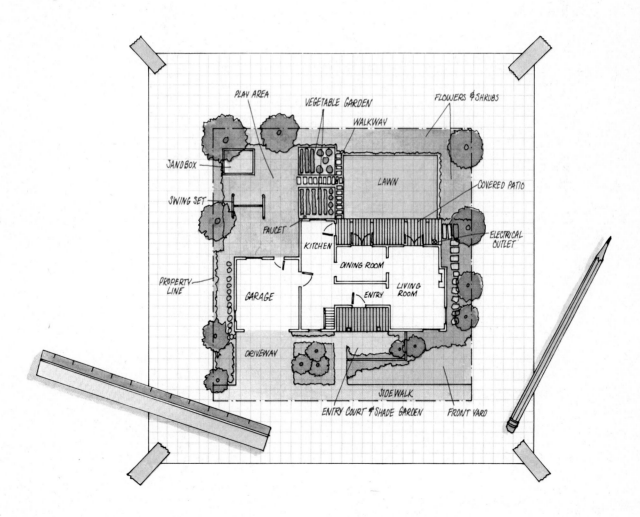

Seven Steps to Successful Storage

3. Identify Activity Areas

Lay a sheet of tracing paper over your plan. With a colored pencil, circle the areas where your family engages in outdoor activities—gardening, playing games, cooking out, sunbathing, working on the car, and so on.

Include activities that occur in your garage or carport if you have one—parking the car, loading the car for trips, doing workbench projects, or even doing the laundry. If your kids play in the garage on rainy days, consider their activities too.

More than one activity is likely to take place in certain spots. For example, you might use your patio mostly for entertaining, but perhaps you occasionally do small woodworking projects there, and maybe the children ride tricycles or rollerskate on it as well.

Similarly, you may do the same activity in more than one place—potting plants, for instance, might occur on the deck, in the garage, and at the entryway, depending on the weather and your mood. Unless you have a convenient central place to store your supplies for such activities and a handy way to

carry them about, chances are the items get scattered over all these locations.

By listing all of the activities that go on in each activity area, you can see where things actually happen, not where everyone thinks they're happening.

Finally, draw lines for the major traffic paths through your yard and to and from the garage, carport, or shed. These lines give you a bird's-eye view of how family members get from one area to another. This will help you keep the traffic pathways clear, open, and direct, so that access to storage areas will be easy and straightforward.

The Activity Areas

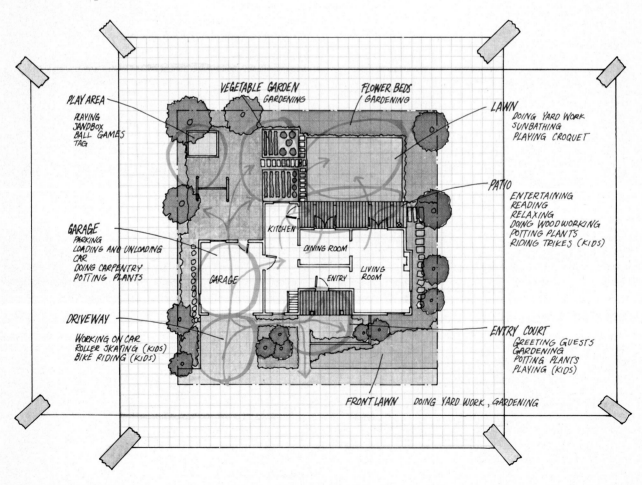

PLAY AREA
PLAYING
SANDBOX
BALL GAMES
TAG

VEGETABLE GARDEN
GARDENING

FLOWER BEDS
GARDENING

LAWN
DOING YARD WORK
SUNBATHING
PLAYING CROQUET

PATIO
ENTERTAINING
READING
RELAXING
DOING WOODWORKING
POTTING PLANTS
RIDING TRIKES (KIDS)

GARAGE
PARKING
LOADING AND UNLOADING
CAR
DOING CARPENTRY
POTTING PLANTS

KITCHEN

DINING ROOM

GARAGE

ENTRY

LIVING ROOM

DRIVEWAY
WORKING ON CAR
ROLLER SKATING (KIDS)
BIKE RIDING (KIDS)

ENTRY COURT
GREETING GUESTS
GARDENING
POTTING PLANTS
PLAYING (KIDS)

FRONT LAWN   DOING YARD WORK, GARDENING

4. Pinpoint the Problem Spots

With your activity map in hand, take a quick walk around your yard. Note on the map all the objects that tend to get left outside, stacked in a corner, or strewn about the yard. These items are additional clues to your living patterns and to problem spots in your current storage situation. They may indicate (1) that some storage locations are too inconvenient to be useful, or (2) that this is where your family actually prefers to use those items, whether or not it's where they're "supposed" to be used. For example, if the shade garden in the entry court is your pride and joy, you may find it annoying to have it constantly cluttered with toys. However, if the kids want to be near you while you're gardening, and if there's no place nearby to put their toys away, they may tend to leave them out when they go back inside with you.

When you've noted the problem spots on your site plan, find a comfortable spot to sit, and study the situation you've just mapped out. At the side of the overlay page, list issues you want to resolve in your new storage plan: "Need a place for the kids' toys near the entry garden." "Decide where the main potting area should be, with room for pots and soil sacks." "Decide whether or not to move the swing set."

At this point, do some thinking about how you want an area to look. For example, you may want the patio kept free of clutter and storage components, so it's always attractive for entertaining; but it may be totally inconvenient to keep the barbeque in the garage. When conflicts between practicality and aesthetics arise, decide which is more important and note that priority.

The Problem Spots

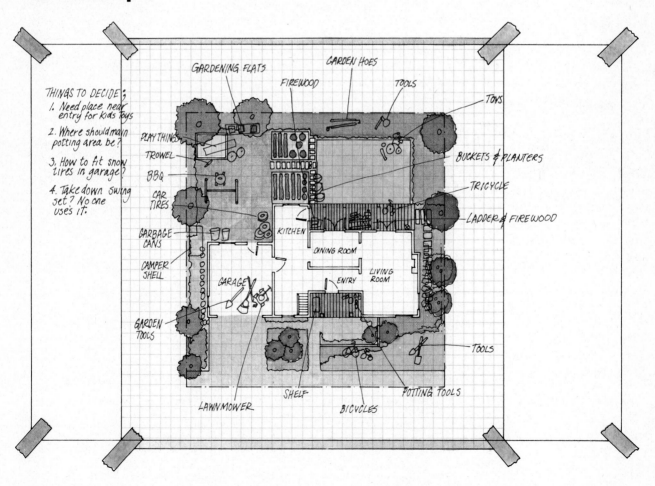

Seven Steps to Successful Storage

5. Assign Storage Locations

With a clearer picture of your problem spots and priorities, take a closer look at existing and potential storage areas. Always begin with your existing storage space. Anything that can be done to increase its effectiveness is likely to take less time, effort, and money than building a new structure.

Make a sketch of that space, and note any feature that can't be changed or that has top priority. For example, if there's no other place to put the washer and dryer except the carport utility closet, assign them that location on your sketch. If keeping your car in the garage is important, show it parked there—with the doors swung open—so you'll know about how much space is left for storage uses.

Next, look at the piles you created at the end of the sort-and-toss process and list the categories of items you have. Starting with the categories that you use most often or in your most important activities, find the best location, vicinity, or space for each. The items should be convenient to the area where you use them, and the location should permit easy access and traffic flow.

Use large, medium, and small circles to indicate about how much room each category will need. Remember that you have wall space as well as floor space, even though it doesn't show in this overhead view, and that this space can hold a lot. Don't try to figure out all the details; make your best guess and move on, until all the categories are designated on your map with circles.

Consider Adding a Storage Structure

Sometimes there are obvious discrepancies between the available room and the bulk of goods that need to be stored in it. No one can jam three bicycles, a lawnmower, woodworking equipment, and garden tools into one small cupboard. Or even if everything would fit in the space you have, you may decide that you really want a workbench, and that you could accommodate one if you had a separate shed. Make your best guesses about space and priorities for existing storage areas. Then, if it seems that a new structure will give you definite benefits consider adding one.

Use your site plan and overlays to play with various locations. Think about the type of structure you want, too. You don't have to install a big shed if that's not what will work best for you. Pages 88–93 will give you some good ideas. Knowing what types of structures you'd like may expand the possibilities for their locations.

New Storage Locations

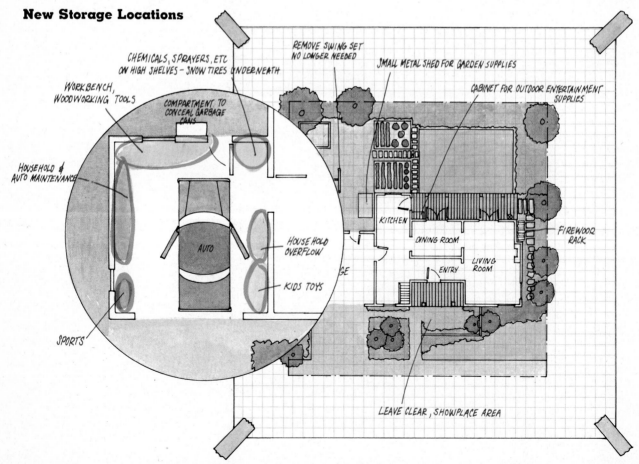

CHEMICALS, SPRAYERS, ETC ON HIGH SHELVES – SNOW TIRES UNDERNEATH

REMOVE SWING SET NO LONGER NEEDED

SMALL METAL SHED FOR GARDEN SUPPLIES

WORKBENCH, WOODWORKING TOOLS

CABINET FOR OUTDOOR ENTERTAINMENT SUPPLIES

COMPARTMENT TO CONCEAL GARBAGE CANS

HOUSEHOLD & AUTO MAINTENANCE

KITCHEN

FIREWOOD RACK

AUTO

HOUSEHOLD OVERFLOW

DINING ROOM

LIVING ROOM

KIDS TOYS

ENTRY

SPORTS

LEAVE CLEAR, SHOWPLACE AREA

6. Match Items with Storage Components

With general storage areas assigned to categories of belongings, figure out the best way to outfit each area so that individual items will stay organized, handy, and neat. You may be surprised to discover that you have definite ideas about how to store almost everything you own: "I want woodworking tools on a pegboard over my workbench; sandpaper in a drawer; nails, screws, and so on in compartments in a top drawer. I want snow tires side by side and concealed, the camper hood hung on the wall or standing on edge on the floor," and so on. It may turn out that you've been devising a storage system for years; you just need to bring your ideas up and out.

Start by sorting the items in each of your categories according to size and other similarities. Make lists of these groupings. Then, next to each item, write down your own idea of the ideal way to store it.

Some items don't suggest obvious storage solutions. Small hand tools, for instance, might be kept in a drawer, on a shelf, on the wall, or in a tool box. Choose an option by matching the way you use the tools with the access and convenience a particular method will give you. If you tend to leave your trowel in the garden, for example, it might make sense to get a carry-all tote for your small garden tools, take it with you when you're working, and then store the tote on a shelf. If you do a lot of work at your potting bench, the tools may be handier hung on a pegboard over the bench.

Use the rest of this book to get an overview of available storage components—and pay particular attention to those you can buy, pictured in the next chapter. You may be surprised at how versatile, attractive, and ingenious they are.

For purposes of planning at this stage, just decide what types of components you want—boxes, bins, drawers, hooks, etc. Whether you'll buy or build these components can be deter-

mined later. Finish matching types of storage components to each group of items. Then make a list of the types and number of components you want for each of your categories.

Basic Storage Methods

When you don't have immediate ideas, the item itself and the way it's used often suggest an appropriate storage method. A riding mower, for example, is heavy, obviously needs to sit on the floor, and should have a clear path to the doorway. The basic storage options are:

■ **Standing on the floor.** This is the best choice for anything very large and heavy.

■ **Raised off the floor.** When an item is too big for a shelf but shouldn't sit directly on a potentially damp floor, you can raise it a bit (see page 46). Firewood and lawn furniture cushions are two candidates for this treatment.

■ **Hanging on a wall.** Almost any object that is either flat or skinny can be stored neatly on the wall with the proper hanger.

■ **Housed in a cabinet.** Cabinets keep things out of sight and away from dust and dirt. A lockable cabinet offers extra security.

■ **Contained in a bin, box, or drawer.** Containers keep small and medium-sized things organized, especially if the container is divided into compartments.

■ **Sitting across roof-ties or joists.** If your shed or garage has exposed ties or joists, they might be a good spot for big, lightweight items that you use infrequently.

Seven Steps to Successful Storage

7. Integrate and Implement Your Plan

To complete your new storage plan, you need to combine the individual storage components into a sensible system. Basically, this amounts to making sketches of each wall until you hit on an arrangement that feels right to you. There's no "trick" to it. You just use your imagination and common sense.

Sketching Floor and Wall Space

Any empty structure is a container, and you want to make full use of all the surfaces—horizontal, vertical, and overhead. So, start by measuring the dimensions of the garage, shed, or other storage place you're working on. Then draw both a floor plan and an elevation of each wall. For the most accurate planning make your sketch to scale on graph paper.

Deciding to Buy, Build, or Hire

To bring your plan to completion and carry it out, you have three options: You can buy the storage components and structures you need, you can build them yourself, or you can hire someone to do the construction for you. Weigh the alternatives, and choose your approach based on the time, money, and skills you have.

Buy. Buying is the fastest and simplest route to a completed storage system. Often it's the least expensive, too. A prefabricated shed, for instance, can generally be installed for less money than one built from scratch. If your time and interest are limited or if you don't have construction skills, this is a practical approach.

Build. Building your own storage projects gives you the opportunity to tailor their size and appearance to fit your home. You don't

Include door and window openings, electrical and water outlets, and any other features that affect how you'll use the space. This will give you a sense of the structure's spatial capacity and will remind you not to block access to mechanical outlets with storage components. If the studs are exposed, indicate them, so you'll know where

things can be easily mounted.

When you're done, you'll have several sketches, each showing part of a principal storage area. It's a good idea to label them, so you can tell at a glance which shows what.

On the drawings for each wall roughly indicate how much space you allotted to each category on your site plan—for ex-

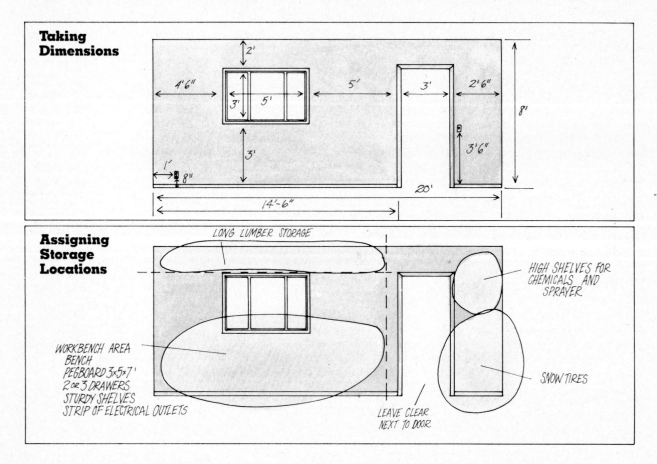

Taking Dimensions

Assigning Storage Locations

LONG LUMBER STORAGE

HIGH SHELVES FOR CHEMICALS AND SPRAYER

WORKBENCH AREA
BENCH
PEGBOARD 3×5×7'
2 OR 3 DRAWERS
STURDY SHELVES
STRIP OF ELECTRICAL OUTLETS

SNOW TIRES

LEAVE CLEAR NEXT TO DOOR

have to settle for someone's standardized product in your non-standard situation. You'll gain flexibility, a more customized look, and the pleasure of installing the product yourself, but you'll spend more time, effort, and money.

Hire. Hiring someone to do the work allows you to have a custom-crafted solution even if you don't have the time or skills to do the construction yourself. Working with a professional architect, contractor, or carpenter (depending on the scope of your project), you'll gain the advantage of having guidance during the planning phase as well. This option is the most costly of the three, but it may be worth it to you. Even if you don't plan to stay in your present home, handsome and well-constructed storage facilities may be a worthwhile investment that will improve your home environment for future users as well as present ones.

erty, follow the same process for its interior area: look at the categories of things you want to store in it and the types of storage components you want. Measure the dimensions of the structure you're considering, and make a floor plan and wall elevation. Sketch various component configurations, determine whether the new structure will work for you, and decide what accessories you'd need to buy.

If your budget is low and your plan extensive, set priorities, and start with number one. Even if time and money constraints limit how much you can do at one time, your plan will provide a foundation for phasing in each stage.

In the next four chapters, you'll find plenty of ideas for storage structures you can buy or build. Refer to them often in your planning. They'll help you extend the order of your household to all of your belongings and all of your space.

ample, about eight feet of wall space for your workbench and three feet for snow tires. Feel free to change these space allotments as you get more and more specific in your planning.

Next, pick a wall to start with. Lay tracing paper over the elevation you made, and sketch out ways to outfit the space. When you're satisfied with arrangements for all surfaces in your existing storage areas, finalize and refine the size and dimensions of the storage units according to the actual dimensions of your space. Then make a shopping list of the things you need in order to carry out your plan.

If you intend to build a new storage structure on your prop-

Finalizing the Scheme

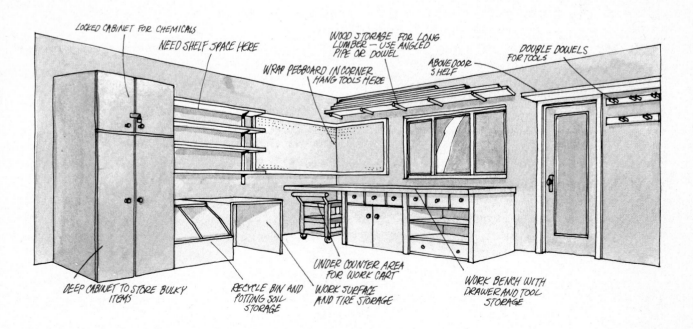

LOCKED CABINET FOR CHEMICALS
NEED SHELF SPACE HERE
WRAP PEGBOARD IN CORNER HANG TOOLS HERE
WOOD STORAGE FOR LONG LUMBER — USE ANGLED PIPE OR DOWEL
ABOVE DOOR SHELF
DOUBLE DOWELS FOR TOOLS
DEEP CABINET TO STORE BULKY ITEMS
RECYCLE BIN AND POTTING SOIL STORAGE
UNDER COUNTER AREA FOR WORK CART
WORK SURFACE AND TIRE STORAGE
WORK BENCH WITH DRAWER AND TOOL STORAGE

Storage Components You Can Buy

You might find the answer to your storage problem at your neighborhood home center or hardware store. Here is a selection of products you can buy to make a garage or shed hold more and work more effectively.

Many people who wouldn't conceive of a kitchen without shelves or cabinets somehow expect that a garage or shed could do its storage job without a similar kind of help. Your outdoor storage space will work much better if you fit it with well chosen holders and containers. It will accommodate more belongings, keep them in better order, and make them easier to use. This chapter gives you a photographic overview of storage aids you can buy—the quickest, simplest, and often least expensive way to improve your storage system.

The box on page 21 tells about the basic ways an object can be stored: It can stand on the floor, be raised above the floor on supports, sit across joists or rafters, hang from a hook, sit on a shelf, be housed in a cabinet, or be contained in a bin, box, or drawer. You'll need to provide the floor or the rafters—the other elements can all be purchased.

Which storage method works best depends on the nature of the object being stored. Tiny objects, such as screws and nails, that tend to come in multiples, are best sorted into suitably sized bins and compartmented drawers. Long-handled tools, such as rakes and shovels, hang neatly against walls and won't get snarled or trip you. Toxic chemicals are safest in a locked cabinet. The most successful storage systems are likely to incorporate every storage method. A number of useful components are shown on the following pages as examples of the variety available on the market. There are single components to store a wide range of items and versatile combination systems to organize various collections of tools, toys, and other possessions.

◀ Mounted above a potting bench, white wire shelves organize gardening and plant care supplies in one convenient place. Widely sold and quickly installed, these and other manufactured products can be arranged to fit out your storage space. From large-scale units like cabinets to detail helpers like these adhesive-backed hooks, you can create good solutions to almost any storage problem using ready-made components.

You may feel that appearance is not as important for outdoor storage as it is inside your home, but if your setup turns out to be attractive as well as practical you'll find it that much more satisfactory. Storage components in your garage or shed don't have to be drab to be functional, as the clean lines and cheerful colors of these products show. To provide a good background for them and give your space a quick uplift, paint the interior white. Not only will it look cleaner and more pleasant, but also, because white reflects light and increases visibility, the space will be safer and more convenient to use.

Appearance is just one factor to consider when buying storage elements. Other key questions to ask are:

▪ How durable is the product? How will it stand up under the use you're planning to give it?

▪ How stable is it? Can it bear sufficient weight for your purpose?

▪ How easy is it to keep clean?

▪ Can it be repaired if necessary?

▪ Does it need to be assembled or installed? If so, how difficult is that process? Is the required hardware provided?

▪ Is it worth the cost?

Use your imagination as you shop. A quick browse through a well-stocked hardware store or housewares department will give you lots of ideas. In addition to utility storage products, some containers and shelves intended for kitchens, bathrooms, or closets may adapt nicely to outdoor storage—you'll discover a variety of possibilities. Even products not meant for storage at all might inspire you. A hammock, for instance, could be slung from the garage joists to hold bulky objects such as sleeping bags and tents. Any item that can support or contain other objects is a potential component for your storage system. Purchasing components will give your storage system a fast start in moving from plan to reality.

Hooks and Racks

A bar fitted with sliding hooks and clips turns a blank wall into functional storage space, as shown in the photo below. The springy, hinged clips hold the long handles of brooms and rakes flat against the wall—keeping them neater and safer than just leaning them into a corner. The movable hangers can be positioned at any point along the bar to accommodate objects of various sizes. Fixed, oversize hooks at either end add to the unit's total capacity.

Magnetic hooks and clips (page 27, top left) cling to steel surfaces; they are especially useful in a metal shed, where options for storage components are limited. They are not designed for heavy weights, but are good for keeping small, easily lost objects in view and in reach. Next to a workbench, the clips can keep project plans out of your way but easy to see while you're working.

Hangers designed for masonry walls (page 27, top middle) let you make good storage use of tricky mounting surfaces, such as concrete or cinderblock, where ordinary hooks can't fasten to the wall with enough enduring strength. These plastic hooks are mounted on spikes that you drive with a hammer directly into the wall.

Tension-loaded clips (page 27, top right) grip objects securely; the items snap out when you're ready to use them. The metal clips come in several sizes to hold objects as thin as fishing rods and as fat as flashlights. The white plastic ones, smaller in scale, can be used on Pegboard as well as a wall.

Multiple-tool holders (page 27, center) offer compact, convenient storage for a collection of hand tools. At left in the photo, a magnetic bar holds the tools securely and neatly, yet taking them down and putting them back could not be simpler. The red tool holder on the right can be hung horizontally or vertically or even placed in a drawer. Its tension corrugations grasp tools tightly no matter how you position it.

Heavy-duty hooks and racks (page 27, bottom) free up floor space by moving bulky objects to your walls. In this photo a bike frame hangs from a steel hook with a long L-shaped arm; a vinyl coating cushions the arm surface and protects the bike from chips and scratches. A pair of such hooks might also accommodate an extension ladder. A smaller U-hook holds the bike's disengaged front tire. For auto tires, an orderly solution could be an industrial-style wall rack.

On this storage bar, two types of hangers slide into place to hold both long and short implements.

Magnetic holders give small, light-weight objects a home on a steel shed wall or cabinet door.

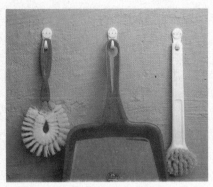

Nailed straight into concrete or masonry, these hooks make it possible to hang implements on walls where other holders won't work.

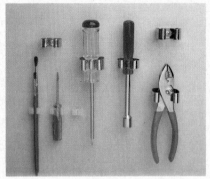

Objects snap in and out of these tension clips which are available in varying sizes.

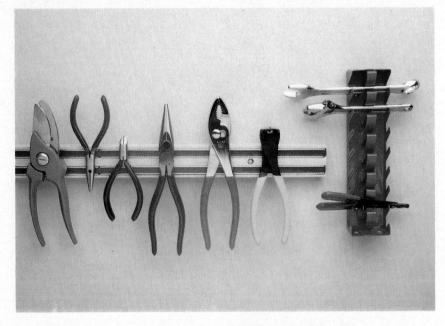

Little tools all in a row stay handy and neat. The holder at left is magnetic. The one at right has tension springs to clip tools in place.

Often awkward and in the way on the floor, big items can take advantage of wall space with hooks and racks of appropriate size and design.

Cartons, Caddies, and Cabinets

Portable boxes and cartons (below left) add carryall convenience to storage, making chores easier and stored items handier to use. The high sides and hand-grips on the plastic crates reduce the risk of dropping awkward objects such as wine bottles and flowerpots, while the open fretwork walls provide good ventilation and an easy view of what's inside. Storage boxes of heavy-duty corrugated cardboard give inexpensive protection to out-of-season sports gear and clothing or to inactive family files. The handholds allow for air circulation as well as ease in moving

the boxes. Moisture can damage the cardboard, so use these cartons where they'll stay dry.

Tool caddies (below right), stocked and ready to go, are indispensable to any home. They can be kept at the workbench for everyday use, yet are easy to carry to on-site tasks. The plastic carriers shown below—lightweight, colorful, and easy to keep clean—organize tools and supplies for household maintenance but could be used just as well for car care, gardening, or hobbies. The ones at center and bottom have partitioned drawers to keep tiny items separated.

The top one expands its capacity with a removable tray. The red wings flip up to close the case and keep its contents secure and dust free.

Cabinets (facing page) are containers on a grand scale. Because they close, hiding the whatnot inside, they give an impression of order and visual calm. Steel models are practical choices for sheds or garages that have no moisture problems; they come in bright colors to cheer up your space. The blue one at left combines the security of a lockable cupboard with the convenience of easy-access shelves.

Easy-to-carry boxes and crates are a good choice for household-overflow storage—items stored outside the living area but used inside.

Portable storage containers, such as these tool holders, gather supplies for an activity and make them easy to use and put away.

Just shut the door, and a cabinet gives you an orderly appearance by screening clutter from view. One that locks is a safe repository for garden chemicals and other toxic substances.

Bins, Drawers, and Trays

Stackable bins and drawers let you spread "up" instead of out, to make good use of tight space. Shown at right are a few examples. The self-stacking white containers are commodious, yet not so deep that small supplies at the bottom are hard to reach. The open fronts allow you to remove things and put them back without taking apart the stack. The smaller red components shown in two sizes, come several bins to a package. A mounting rack is included. You can hang them on a wall or stack them on a workbench or shelf. Two drawers are better than one: A kit of uprights connects a pair of rollout units to make an interior cabinet space hold more with better access.

Containers with compartments are best for tiny hardware items, hobby parts, and little tools. Flat, divided trays can organize a drawer or shelf. Some of the partitions in the white trays shown at top of page 31 are movable, to let you create sections of appropriate size. Veterans of many well-ordered workshops, the mini-cabinets in the same photo, separate supplies into small drawers made of clear plastic so the contents are easy to see. (If you prefer, you can apply gummed labels to tell you what's inside.) The cabinets can be stacked, hung on a wall, or hooked to a Pegboard surface. The larger, steel-framed blue one has a convenient carrying handle.

The space-saving components shown at the bottom of page 31 create efficient storage in otherwise wasted spots. The wall-mounted unit at upper left takes little room; the bins can be sectioned with dividers, and they rotate out 180° for easy access. The two-tiered blue holder was put together with modular trays and supports; more layers could be added. It can tuck into a shelf corner, attach to a wall, or hang

from pegboard hooks. The brown bins, made of high-impact plastic, come with clips for mounting to a Pegboard, wall, or cabinet door. Their lipped edges keep bottles from spilling and hold round objects.

Under a shelf or work surface, try one of the space expanders shown at the bottom of the

photo. The tidy drawer on the left keeps dust and dirt away from its contents; it glides out on steel slides. The vinyl-covered wire basket on the right offers visibility and good ventilation for items stored there. Some styles come with clips so that you can hang additional baskets below the first.

If the items you want to store won't stack neatly, stack the storage components instead. Many varieties permit you to make stylish and efficient use of space.

"A place for everything and every-thing in its place"—it's readily accomplished when you have lots of compartments.

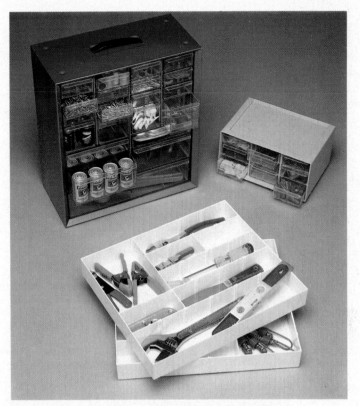

Small storage aids such as these make effective use of space on walls or under shelves and workbenches.

Shelf Systems

Metal shelving, rugged and durable, is a storage mainstay in a dry garage or shed. The freestanding units below can be positioned against walls or project out into the space, creating aisles. If the walls are masonry or metal, freestanding shelf units are the simplest way to expand the area's storage potential.

Though the steel mesh shelves on the left look airy, they are extremely solid and stable. Because they were originally designed for commercial use, they can support considerable weight. They are also easy to maintain, since dust and dirt don't build up on them. You can tailor a unit to fit your space and needs, because this shelving comes in a wide range of heights, lengths, and depths. Accessories include shelf edging and dividers.

The well-braced steel utility shelving on the right is a strong, economical workhorse. Once available in gray only, these units are now manufactured in a selection of colors to brighten your storage area.

Versatile, small-scale shelf units (page 33, bottom) come in styles to serve almost any space and purpose. The rolling red cart could be a portable toy box, collecting playthings strewn about the house and yard. Or it might be an outdoor entertainment center, storing barbecue and party supplies in the shed or garage and trundling them out to the patio for use. The beige molded plastic unit keeps cleanup supplies handy on the inside of a cabinet door, and the white vinyl-covered wire shelves might put the back of a regular door to work as well. The lipped edges on both pieces keep stored items from tumbling off as the doors open and shut.

A flexible system of components lets you transform a little

Sturdy metal shelving can handle large, bulky items well. Before you buy a set, make sure it will hold the weight you want it to bear.

nook or an entire wall into an organized, active storage system. The one in the photo to the right has upright tracks that can be fitted with bracket arms for shelves or with crossbars to support hooks and hangers. The elements can be arranged in whatever configuration works best for your wall space and possessions. To expand its potential further, sit bins or baskets on the shelves to hold small objects. The result is a storage center where all of the gear for an activity—whether big, little, or bulky—is stored conveniently together, as the camping and fishing equipment is here.

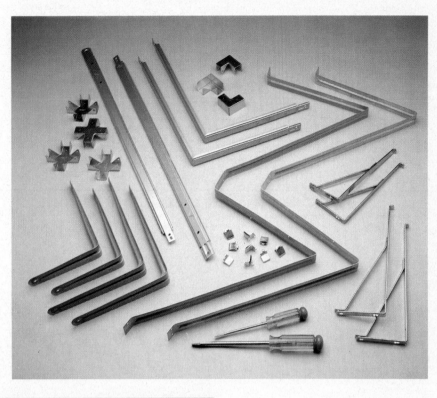

A selection of hardware components, plus your choice of lumber, allows you to build shelving quickly and inexpensively.

These lightweight shelf units are designed to move without spilling their contents. The wheeled cart at left shows that shelving can be portable, while the other two units turn doors into swing-open storage spaces.

Combination Systems

Putting up do-it-yourself shelves is a project that pays big storage dividends for a small investment of time, skill, and money. The still-life arrangement below shows three basic types of installation hardware. Brackets with projecting arms screw onto a wall stud; the brackets have nail or screw holes so you can securely fasten the shelf to them. Straight shelf tracks affixed to a wall or cabinet support clips or brackets on which the shelves rest; the shelves are adjustable in this system (see page 45 for mounting instructions). The little clips shaped like *X*s and *L*s join boards to form corners; screws hold them in place. The clips are decorative as well as functional.

A wall-mounted grid (page 35, top) of vinyl-covered wire uses an assortment of accessories to form a storage center that's both handy and handsome. It prevents roller skates and wet raingear from dirtying the main part of the house. Hooks, rings, and a basket keep the paraphernalia high and drying. Other components can turn the grid into an organizer for garden supplies or workbench tools.

The stand of sliding wire baskets (page 35, bottom) is just one representative of a collection of products that can be mixed and matched for systematic storage. With a wide variety of components to choose from—shelves and baskets, wall-mounts and freestanding pieces—you can create an effective, versatile storage system. First you'll want to plan your space and think through your needs carefully. The preceding chapter will assist you in that process.

Sold as parts of a coordinated system, these tracks, hoods, bars, and brackets combine into a storage center for outdoor recreation gear.

The basket on this grid lets you hang items that you wouldn't ordinarily store on a wall. More traditional hooks expand the system's ability to accommodate a variety of items.

Shelves and other components can be joined with this basket stand to form an effective system for storing a broad range of belongings.

Combination Systems

Pegboard organizers can create a comprehensive storage system, incorporating an almost endless array of hangers (shown below), shelves, bins, and other containers. You can use them to make a storage wall that will keep all manner of tiny to medium-size objects neat, visible, and handy. Tool racks and hooks, for example, hanging over a workbench, keep screwdrivers and hammers within easy reach of the resident carpenter. At home on its designated hook, the flashlight can be found without a hunt when the next emergency occurs.

Colorful plastic bins and racks (page 37, top) make Pegboard a rainbow of storage—good-looking yet functional. Small, sectioned bins in various configurations keep washers and screws separated and easy to grab when needed. The black rack has slots for hanging narrow-bladed tools, and the red tray holds bottles securely in its molded, cup-shaped base. A number of the components shown on previous pages in this chapter are also designed to be used on Pegboard.

Red-capped jars (page 37, bottom) store nails, pushpins, and the like on Pegboard with no danger that they'll spill. Prongs on the caps hook into the Pegboard holes; the plastic jars can be twisted loose while the caps stay in place, or the whole unit can be unhooked.

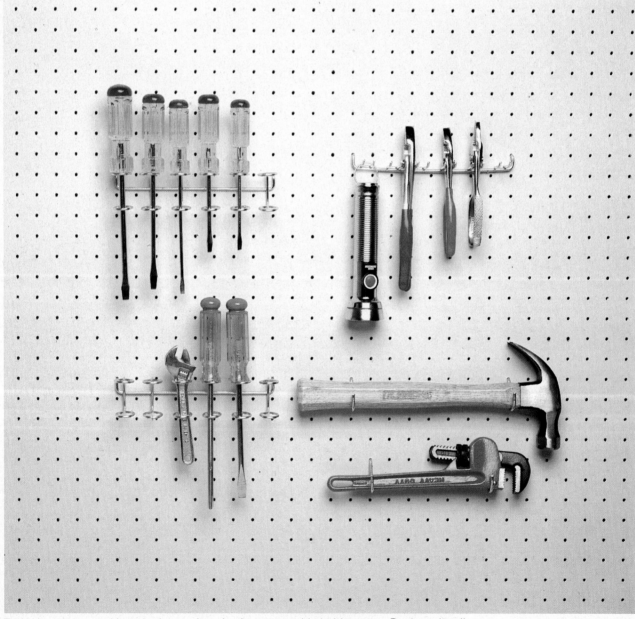

Tools rise above workbench clutter when they're arranged in holders on a Pegboard wall.

Pegboard accessories are not limited to hooks. Bins, racks, and even shelves can make a Pegboard storage system successful.

Lines of jars with brightly colored caps organize tiny objects and hang on a Pegboard surface.

Storage Projects You Can Build

Shelves, cabinets, racks, and bins are the furniture of storage. If you build your own, you can tailor them to your space. These projects can be constructed as shown, or modified as needed, to organize your storage areas.

A little lumber and some nails can go a long way toward helping you put your storage plan into action. While ready-made components and devices may suit some of your needs, for others you may want custom solutions. This chapter contains projects you can build to get more use out of your garage, shed, carport—or even your balcony—by helping you store more in it.

Certain storage needs are hard to meet with store-bought components. For example, you may want help in confining and organizing heavy, bulky objects or pieces of equipment that are to be stored on the floor. Or you might want your storage component to have a built-in look, fitting neatly and precisely into its surroundings as if it had always been there.

Building you own components enables you to customize their dimensions and functions, so that you create a storage solution that works perfectly rather than compromise with a near-fit. You'll get satisfaction from several aspects of the project: from knowing that the craftsmanship is up to your standards, from the pleasure of working with wood, and from the sense of accomplishment that comes from creating something yourself. Some of the devices are simple ones that require only a few minutes' time and a couple of basic tools; others are more complex and take greater time and skill.

Like the components to buy in the previous chapter, the projects described here are arranged according to how they function: some help you store things that need to rest on floors or sit on shelves; some are for hanging things against walls: others help you contain your belongings, or conceal them from immediate view.

Your objective is to make your storage space easy, convenient, and pleasant to use. Consider these three simple ways to further that objective:

■ Paint the whole space and the storage components you build a light, neutral color—even white! The background uniformity will reduce the chaotic look storage spaces often have. White or pale colors aid visibility.

■ Light the space well. It will upgrade the ambiance and will make the space safer and easier to use.

■ Put labels on boxes, shelves, and drawers to help people find things and return them to their proper places.

Some Notes on Materials

Plywood is specified for most of these projects, but particle board could be used as well. Choose high-density (60#) panels the same thickness as the plywood. Particle board is a low-cost material, it has no voids, and its smooth, grain-free surface takes paint well. But it isn't as rigid as plywood over long spans, which means that it tends to sag or bend when loaded. It's also heavier, and it's harder to cut, nail, and screw into.

If the project will be exposed to the weather, construct it with exterior-grade plywood and waterproof adhesives. For any part that touches the earth, use pressure-treated wood. Because this wood has been chemically treated, it isn't subject to decay or rot caused by moisture and insects. However, the chemicals used in pressure treatment are toxic, so the wood should be handled with care.

For more storage project ideas, see Ortho's book, *How to Design & Build Storage Projects*. It gives directions for making a variety of practical components and for combining them into made-to-order systems for both indoor and outdoor storage.

A bank of clean-lined white cabinets along the wall gives this garage a sleek look, unencumbered by the storage clutter that often characterizes such spaces. Home-built projects such as this can be customized to your individual situation. Instructions for cabinets using similar construction techniques appear on page 48. Page 47 gives more ideas for screening your storage from view.

Hooks and Racks

Make Tool Hangers

Doubled-up Nails

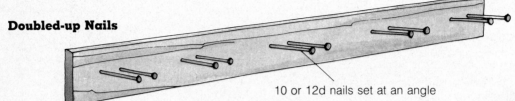

10 or 12d nails set at an angle

Strapping a wall with a board gives you a neat and simple way to hang long-handled tools and other implements. You can double the wall's capacity by stacking two boards one above the other and staggering the hangers.

The hangers can be nails or dowels. They are installed in pairs. For nails, mount 1 by 4 boards to the wall, and drive 10d or 12d nails in at a slight upward slant, far enough into the wood for the nails to be stable. Proceed with a pair of nails for each tool, one pair at a time, to be sure the spacing between the tools is correct.

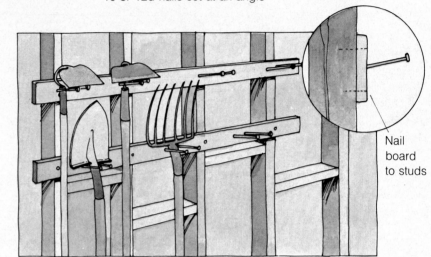

Nail board to studs

Stagger pairs of nails so that one row of tools hangs between the other row of handles

Doubled-up Dowels

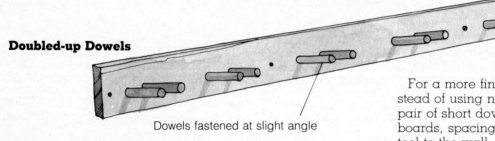

Dowels fastened at slight angle

For a more finished look, instead of using nails, install a pair of short dowels into 1 by 6 boards, spacing them tool by tool to the wall. When drilling holes for the dowels, use a drilling template made from a block of wood to keep all angles the same.

Fastening the boards to the wall. Nail the boards directly to the studs with 8d nails. Boards can be fastened to a masonry wall by drilling holes into the wall with a masonry bit and inserting special expansion sleeves for the mounting bolts.

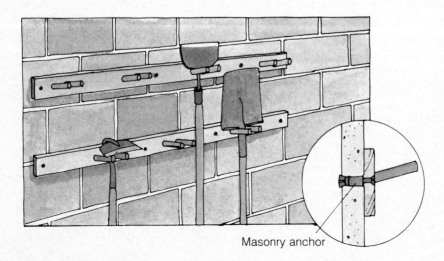

Masonry anchor

Mount Pegboard

Pegboard is convenient for organizing a multitude of little items. To mount it on a flush surface, simply nail 1 by 2 furring strips vertically along each wall stud with 8d or 10d nails. Horizontal strips behind the top and bottom edges of Pegboard are optional but will give it more rigidity. Screw Pegboard to the furring strips.

To give Pegboard a neater appearance, you can make a frame for it instead of using furring strips. To accommodate Pegboard use 2 by 2 lumber and rout a groove in one side of each piece.

Pages 36–37 show some of the hangers and accessories that make Pegboard more versatile for storage.

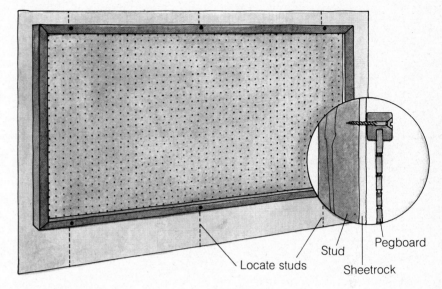

Locate studs
Stud
Pegboard
Sheetrock

Hang a Platform from Joists

A suspended platform uses high, otherwise wasted space for light to moderate loads.

Making the slings. Assemble two slings from 1 by 4 lumber. Apply glue, and lap the joints as shown in the illustration. You can use 6d nails, nailing from both sides, or 8d nails, nailing from one side and clinching the points over in back to help hold the boards in place.

Hanging the structure. To mount the first sling, tack one upright in place with a single nail. Hold a level on the crosspiece to determine the right position for the second upright, and mark the point on the joist. Secure the sling with lag bolts, screws, or nails. Repeat with the second sling, using a level (and a long straightedge, if necessary) to make it level with the first.

The platform can be 3/4-inch plywood or particle board tacked to the crosspieces to make it more stable. Or use the slings without a platform to hold items of appropriate size.

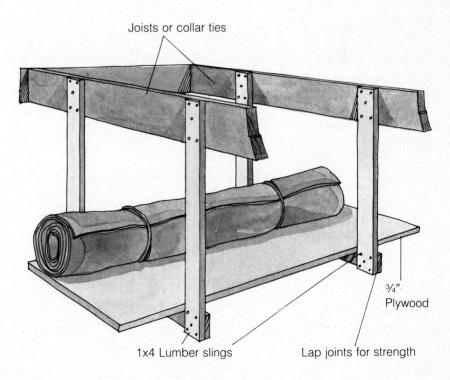

Joists or collar ties
3/4" Plywood
1x4 Lumber slings
Lap joints for strength

Containers

Build Multi-Purpose Bins
Bins such as these can house potting soil sacks, or bottles and cans collected for recycling, and keep them from spilling out all over. The shelf on top braces the structure and provides a bonus storage spot.

To make these bins, use 3/4-inch plywood, with 1/4-inch hardboard or plywood for the back. Cut all the pieces to size. Use 7d box nails to attach the front and top to each side and to the divider. If desired, add a bottom of 1/2-inch plywood, fastened with 6d box nails. Then attach the back with 4d box nails. To make the box stronger, glue the joints before nailing them.

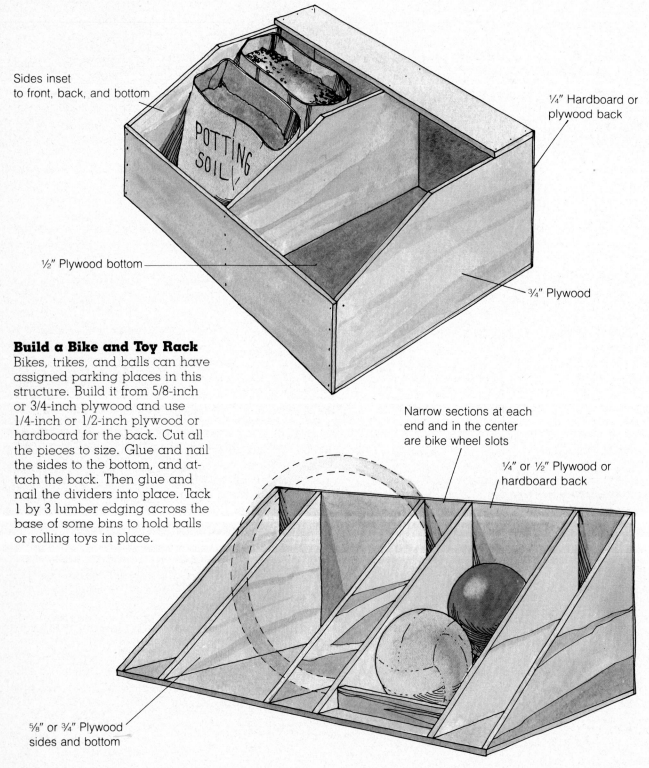

Sides inset
to front, back, and bottom

¼" Hardboard or
plywood back

½" Plywood bottom

¾" Plywood

Narrow sections at each
end and in the center
are bike wheel slots

¼" or ½" Plywood or
hardboard back

⅝" or ¾" Plywood
sides and bottom

Build a Bike and Toy Rack
Bikes, trikes, and balls can have assigned parking places in this structure. Build it from 5/8-inch or 3/4-inch plywood and use 1/4-inch or 1/2-inch plywood or hardboard for the back. Cut all the pieces to size. Glue and nail the sides to the bottom, and attach the back. Then glue and nail the dividers into place. Tack 1 by 3 lumber edging across the base of some bins to hold balls or rolling toys in place.

Build a Lumber Bin

This unit helps you organize all types of wood materials, from plywood sheets to lumber and molding. Simply a variation of the bins pictured on the facing page, this unit's dimensions and number of compartments can vary to meet your own needs. Whenever you're working on a project design, sketch out the various pieces and note the exact dimensions.

The bin's construction. Cut all the pieces to size, using 1/2-inch or 3/4-inch plywood for the sides and partition pieces, 1/4-inch to 1/2-inch plywood or hardboard for the back, and 1/2-inch plywood for the bottom.

Use glue and 7d nails to put the unit together. Assemble the right and left sections separately, by gluing and nailing the sides to the partitions. Nail the bottom to both assembled side sections, then attach the back. Nail on the front pieces last, as they overlap the section sides and the bottom.

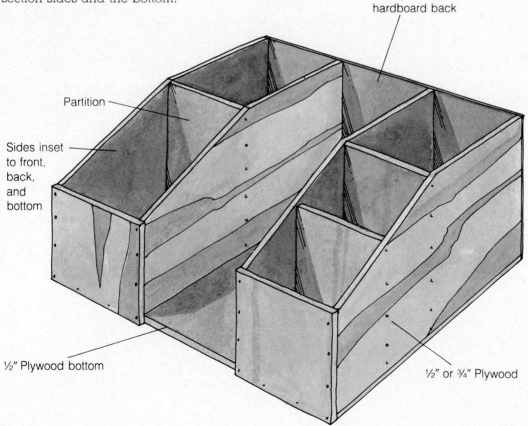

¼" Plywood or hardboard back

Partition

Sides inset to front, back, and bottom

½" Plywood bottom

½" or ¾" Plywood

Shelves

Build a Case of Shelves

A case of wooden shelves can give you a great deal of extra storage capacity, while keeping your things well organized, easily accessible, and neat in appearance. Use 1 by 12 lumber, or, for deeper shelves, rip 3/4-inch plywood to the width you choose. Cut the pieces to length. The shelves should be no longer than 36 inches, or they'll tend to sag under heavy loads.

Mount two shelf tracks to the inside faces of each side piece. If you want to recess the tracks, cut grooves with a router. Otherwise, screw the tracks directly on the surface. Make sure that the tracks are parallel and the slots are level with one another.

Glue and nail the sides to the top and bottom pieces, using four 8d nails at each end. Then attach the two nailing strips with glue and two 8d nails at the top and bottom of the back as shown.

Attach the unit to the wall with 2 1/2-inch wood screws. Hang shelf clips in the tracks, and install the shelves. For a free-standing unit, omit the nailing strips and install a back made of 1/4-inch hardboard or plywood.

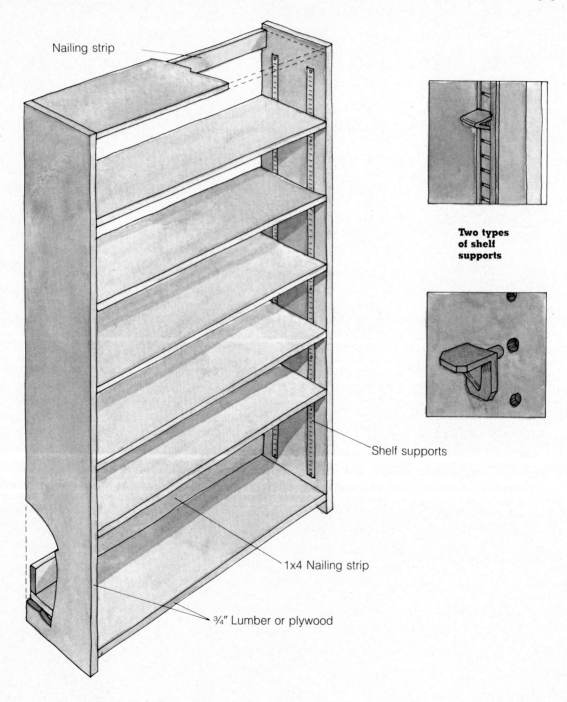

Nailing strip

Shelf supports

1x4 Nailing strip

¾" Lumber or plywood

Two types of shelf supports

Create a System of Shelves

By suspending shelves between two shelf cases, you can create an effective storage wall lined with shelf space. Bamboo shades, mounted at the top of the cases, can be lowered to create an instant impression of order (see page 47).

Construct two shelf units as described above. Before assembling them, install two tracks on the outer side that will face the other shelf unit. Offset the tracks from those on the inside faces, so that the screws and grooves won't collide. Again, take care to make the tracks parallel and level with one another. To fit the middle shelves, first mount the cases to the wall. Then measure the space between the tracks in the middle section. Cut the shelves to this length, and put them in place.

Mount Shelf Track and Brackets

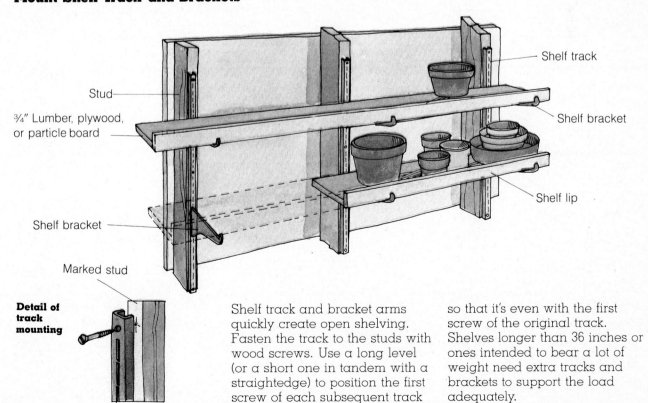

Stud

¾" Lumber, plywood, or particle board

Shelf bracket

Marked stud

Detail of track mounting

Shelf track

Shelf bracket

Shelf lip

Shelf track and bracket arms quickly create open shelving. Fasten the track to the studs with wood screws. Use a long level (or a short one in tandem with a straightedge) to position the first screw of each subsequent track so that it's even with the first screw of the original track. Shelves longer than 36 inches or ones intended to bear a lot of weight need extra tracks and brackets to support the load adequately.

Platforms and Screens

Make a Raised Platform

To protect items from floor moisture and dirt, use 4 by 4s to support a plywood platform. For very heavy loads, set the 4 by 4s no farther apart than 16 inches on center; for lighter loads, 24 inches on center. If the platform is to go outside, rather than in a floored garage or shed, use redwood or pressure-treated 4 by 4s. If the surface should allow air to circulate, lay 1 by 4 boards across the 4 by 4s rather than a sheet of plywood.

Build a Box Platform

A higher platform can be made using 3/4-inch exterior grade plywood. Cut the two sides and top to size, and nail them together with 7d box nails. Cut a back to fit from 1/4-inch hardboard, and nail it to the top and sides with 4d box nails. (The same construction at smaller dimensions is a good way to add a self-supporting shelf to any counter, shelf, or work surface.)

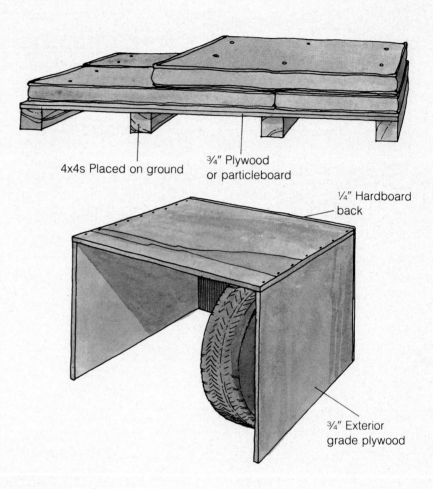

4x4s Placed on ground

¾" Plywood or particleboard

¼" Hardboard back

¾" Exterior grade plywood

Lay a Platform Overhead

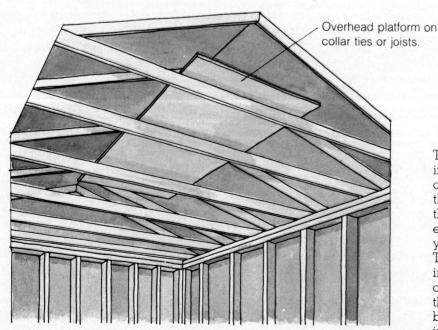

Overhead platform on collar ties or joists.

To make use of overhead space in an open-gable garage, slide a sheet of plywood in to rest on the roof ties. Though you might think that you could obtain easier access or more space if you cut a tie or two, don't do it! They all have a critical function in supporting the roof. This type of storage works best for objects that aren't too heavy, fit easily between ties, and are used infrequently.

Hang Bamboo Shades

A bamboo shade on a case of shelves rolls up quickly to allow you access—and lowers just as fast to present a neat, attractive appearance. Shades are easy to mount: Simply screw two or three hooks to the top edge of the case and hang the shades from them. A row of shades mounted on overhead framing members can screen an entire section of your storage area from view and still allow fast, easy access to the stored items.

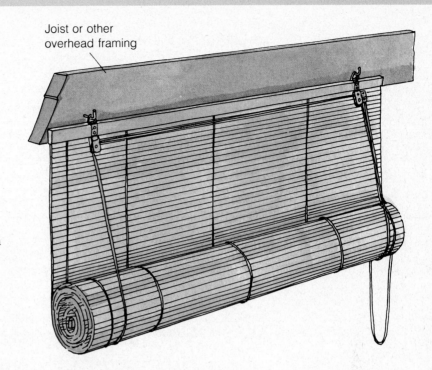

Joist or other overhead framing

Hang a Wall of Sliding Doors

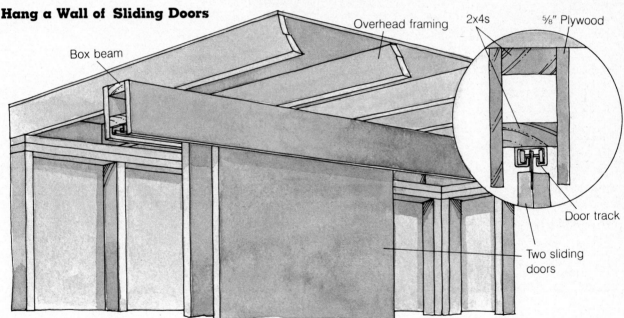

Box beam

Overhead framing

2x4s

⅝" Plywood

Door track

Two sliding doors

Even if they're organized, lots of objects of different sizes, shapes, and colors can present a visual effect of clutter. With a wall of sliding doors, you can section off a storage space and conceal it from view at the same time.

To provide a sturdy mounting surface for the double-channel sliding door track, build a box beam from plywood and 2 by 4s. Start with a 2 by 4 the length of the area you'd like to conceal. Fasten it to the overhead fram-

ing members or to the ceiling it-self. To make sure it is straight, measure out from the adjacent wall, whatever distance you choose, and snap a chalkline. This will guide your placement of the beam as you fasten it in place.

Rip lengths of 5/8-inch plywood for the sides of the box beam. The pieces should be wide enough to fill the gap between the ceiling plane and the top of the roller and track assembly for

the doors. Fasten the door track to the bottom 2 by 4. Glue and nail the plywood side pieces to it with 6d nails, 6 inches on center. Then glue and nail the U-shaped assembly to the top 2 by 4, to form the box beam.

To keep the sliding door panels light and maneuverable, use hollowcore doors or sheets of plywood paneling. Attach the roller brackets to the doors, allowing for floor clearance, and hang them on tracks.

Cabinets

Build an Overhead Cabinet

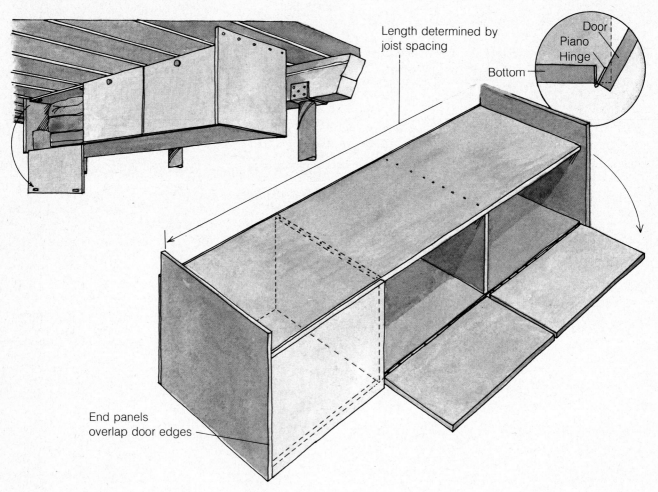

Length determined by joist spacing

Door
Piano
Hinge

Bottom

End panels overlap door edges

This plywood cabinet provides attractive, lockable storage, yet leaves floor space free. A series of these cabinets provides a practical way to add storage to a carport. A series of them mounted under the perimeter of the carport roof can hold a considerable quantity of light- and medium-weight goods (see page 13).

Assembling the cabinet. Because the box hangs from the rafters, the top and bottom panels should measure the distance across the number of rafters the cabinet will span, less 1 1/2 inches—the combined thickness of the sides. Measure from the outside edge of the first one to the outside edge of the last, up to a maximum of 8 feet.

Glue and nail the dividers

between the top and bottom pieces, using 6d ring-shank or 8d box nails. Attach the sides in the same way, making them project above the top. Then nail on the back flush with the top.

Attaching fold-down doors. Cut doors from 3/4-inch plywood, and attach lengths of piano hinge to each door. Hang the doors one at a time. Position the door, mark the screw positions, and fasten the piano hinge to the cabinet. Install magnetic catches (and locks, if desired). You can enclose the space with one long door, but it will be heavier and more awkward to handle.

Using side-mounted doors. Cut two shallow mortises in one side of each door, and attach a leaf

hinge at each one. Screw the other half of the hinge to the inside face of the end panel or divider. Install the catches and locks opposite the hinges.

Hanging the cabinet. Attach the module to the rafter faces using 2-inch by No. 8 wood screws, 4 inches on center. For extra support, screw the top of the box to any rafter edge it touches.

To create a bank of cabinets, make each of the others 3/4-inch shorter than the original one. Hang the first box as described. Butt the end piece of the second box against the end piece of the first, and screw through both projections into the rafter, using 2 3/4-inch by No. 8 wood screws. Fasten the other end directly to its rafter as the first cabinet was fastened.

Build a Floor-Standing Cabinet

This cabinet gives you a convenient counter work surface in addition to storage space. Two cabinets with a work surface installed between them create a practical workbench. Make the unit a comfortable working height for you.

Cut all the pieces from 3/4-inch plywood. Drill holes for the shelf brackets in the end panels and partitions. Place them 2 inches on center and 2 inches in from the front and back edges; measure the spacings carefully to be sure the shelves will be level. In the end panels, drill no deeper than 3/8-inch. In the dividers, drill all the way through to accommodate shelf hangers.

Assemble the case by gluing and nailing first the bottom, then the top, to all four vertical panels, using 6d ring-shank or 8d finish nails, 4 inches on center. Nail the back into place, also 4 inches on center.

To make a kick space, nail the cabinet to lengths of 4 by 4 cut 4 inches shorter than the cabinet depth. If you need to level the unit, shim it up. Screw the back piece to a few wall studs to secure the cabinet.

Mounting a work surface between two cabinets. Build two storage units and carefully install them so that they are level and square.

Screw 2 by 4 nailing cleats across the facing sides of the cabinets, along the back wall, and across the front of the opening. Set the top edge of each cleat below the top edge of the cabinets exactly the thickness of the work surface—whether it is 3/4-inch plywood, butcher block, or other material. Cut the work surface to length; then counterbore the cleats and fasten the work surface with wood screws.

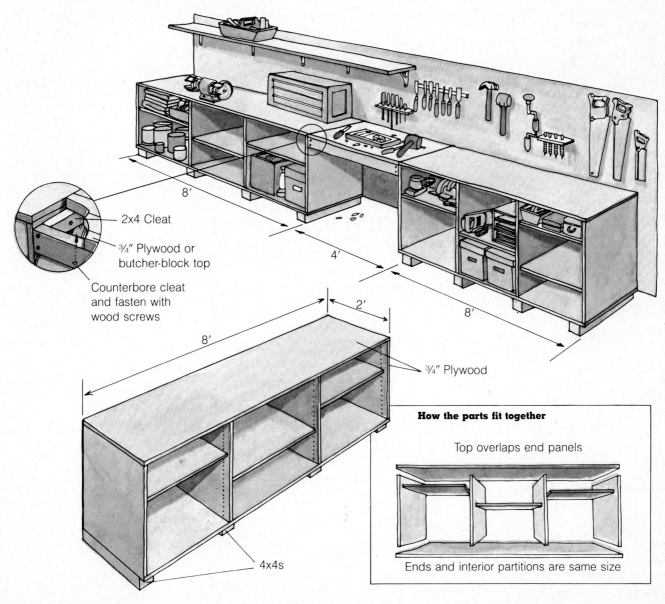

2x4 Cleat

¾" Plywood or butcher-block top

Counterbore cleat and fasten with wood screws

8'

4'

2'

8'

8'

¾" Plywood

4x4s

How the parts fit together

Top overlaps end panels

Ends and interior partitions are same size

Combination Systems

Create a System from Cabinet Modules

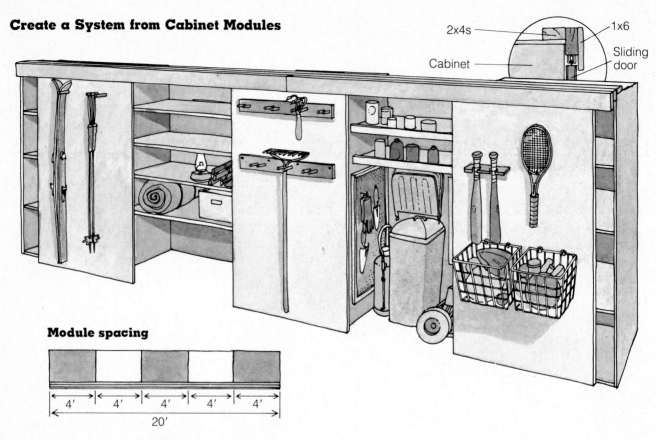

2x4s

1x6

Cabinet

Sliding door

Module spacing

4'   4'   4'   4'   4'

20'

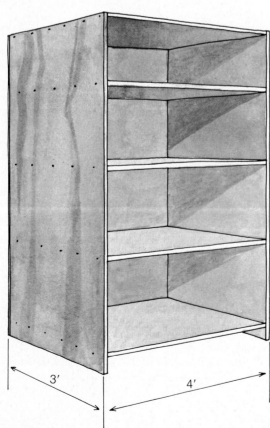

3'   4'

This project combines cabinet, shelves, hooks, hangers, bins—even floor space—into a generous, well-organized storage system. Dual-purpose sliding doors close off the cabinets and provide specialized storage space. As shown, the unit will fill a long wall in a carport or garage. Vary the size and number of the components to make them work best in your situation.

Building the basic case. Cut the cabinet panels and shelves from 3/4-inch plywood. Mark the shelf locations on the inside faces of the two sides, and scribe corresponding nailing lines on the outside faces. Glue and nail one side piece to the top, the bottom, and all the shelves, using 6d ring-shank nails or 8d finish nails. Attach the second side in the same way. Cut a piece of 1/4-inch plywood or hardboard for the back. Glue and nail it in place with 4d box or ring-shank nails.

Installing the modules. Build two or three basic cases, depending on how long you want the storage wall to be. Space them evenly against a wall, and use a level and shims to make each one plumb. Secure them to the studs with three or four wood screws.

Cut shelves as desired for the spaces between cabinets. If you want them to be adjustable, measure, mark, and mount tracks on the facing cabinet sides. Otherwise, mark shelf locations on the cabinet walls, measuring the placements carefully so that the shelves will be level. Glue and nail the shelves in place with 6d ring-shank nails or 8d finish nails.

Mounting the track and door.
To support the door track, you'll need two 2 by 4s, each the length of the total unit assembly. Nail them together at right angles with 16d nails, 16 inches on center. Screw the door track to the bottom edge of the face 2 by 4. Position the 2 by 4 assembly on the top edge of the cabinets, and drill pilot holes to fasten it to the cabinets. Attach it to the front edge of the cabinets with 3-inch by No. 6 screws, then screw up through the cabinet tops into the top 2 by 4

using 2-inch by No. 8 screws. If you want to conceal the door track from view, nail a 1 by 6 board to the front 2 by 4, keeping the top edges flush.

Make doors out of 3/4-inch lumbercore plywood, which stays flat and doesn't warp. Or use 1/4-inch plywood or hardboard with a frame of 1 by 4 lumber on the front or back for bracing. Premade doors can be used if you can find the right size for your units. Attach roller brackets to the doors, and mount the pegs, pockets, or other storage devices you've decided to use on the doors. Hang the doors in the track.

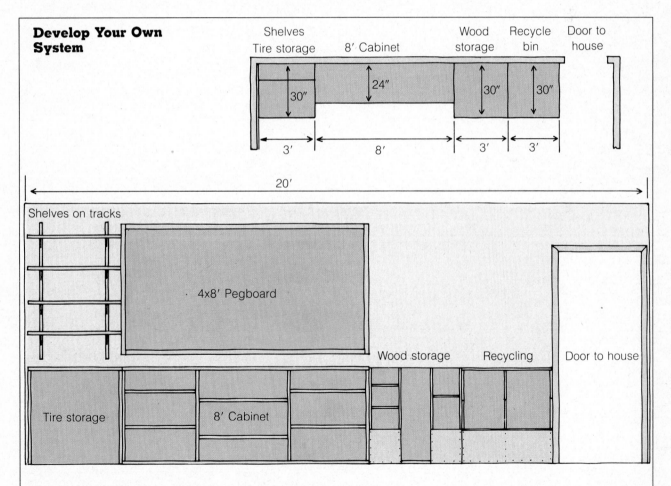

Develop Your Own System

The storage wall shown is only one of many possible ways to combine the ideas in this chapter into a functional storage area. This arrangement brings together the shelves on tracks (page 44), a box platform (page 46), a floor-standing cabinet (page 49), a scrap-lumber holder (page 43), and bins (page 42). The tire storage platform, cabinets, and bins have been built to the same height and depth to create a smooth, unified appearance. Centering the 8-foot Pegboard panel over the 8-foot cabinet maintains a pleasant visual order, too.

Prefab Sheds: Structures You Can Buy

If you need a new storage structure, prefab sheds and construction kits can simplify the building process. Should it be wood or metal, large or small? How can it be blended into the yard? See the options presented in this chapter.

The storage plan you developed in the first chapter may call for a new outdoor structure to supplement your existing space. This chapter provides you with an overview of the types of storage structures you can buy.

Most people picture sheds when they think of outdoor storage, but your plan may suggest other possibilities—a bike locker, a firewood shelter, a container to hide your trash cans, or a small compartment that tucks into a limited space. To help you determine the appropriate structure, its size, and the features you want it to have, think about how you plan to use it. For example, if potting activities will share space with storage, you'll want a window for light and ventilation. If the structure will house just a few large, bulky objects, its capacity for shelves and hooks won't matter. On the other hand, if you plan to store lots of little items there, you'll want to be sure such components can easily be added to the structure.

If you decide to add a storage shed to your property, essentially you have two options. You can buy your future shed in the form of a prefabricated kit or build it from scratch. The difference in the amount of work required isn't always significant, however. Prefabricated sheds are not ready-to-use, finished products; they're actually projects (see page 55). You'll return from the store with a stack of materials or a package of parts, plus instructions for putting them together. The amount of labor, skills, and tools required can vary greatly,

depending on the kit you purchase. Shed kits, both metal and wood types, are likely to be less expensive to construct, however, than comparable structures you build entirely from scratch.

If you decide to buy a prefabricated shed, you face another fundamental choice: Should it be metal or wood? This decision is important; each material has specific functional and aesthetic implications that you'll want to consider. To help you decide which type of shed will meet your needs, the chart on the next page compares them. Other factors you'll want to take into account include:

■ **Codes and permits.** Check with your local building department before erecting any structure.

■ **Cost.** Prices range roughly from $100 to $700, with wood sheds costing a bit more than metal ones. Site preparation and any materials not specifically included in the package (such as anchors for metal sheds) will be extra.

■ **Appearance.** Any new structure in your yard will be part of your life for a long time. To have it blend most congenially with your home's surroundings, you need to match the shed's styling, color, and, in the case of a wood shed, finish materials with the existing house and yard.

■ **Siting.** The site you choose will affect how large your shed can be, how easy it will be to erect, how convenient it will be to use, and how durable it will be. The discussion of siting on page 67 will help you determine the best spot to put the shed.

■ **Climate.** All kinds of weather affect a building (see page 68), so pick a structural design that is appropriate for your climate. A metal shed in particular is vulnerable to damage from strong winds, heavy snows, or salt air. You must tie it down securely, keep snow and ice from accumulating on the roof, and choose aluminum rather than steel for a seaside location.

◀ This metal shed was tucked into a side yard, where it's out of the way yet accessible. A bed of gravel underneath keeps it well-drained. The tall fence hides the shed from the front yard and street, but its openwork top maintains an airy look. To make the shed blend in still more, it was repainted solid white to match the house and fence. Periodic painting helps keep a metal shed in good shape.

First Decision: Wood or Metal?

The right shed to buy is the one that works best for you on three levels: It fits your overall storage plan; it requires the amount of time, effort, and money you choose to invest; and its structural design and appearance are appropriate for your site. Those criteria will point directly to a metal shed for some households. For others, a wood shed will be the best choice.

Metal and wood materials differ in their structural capabilities and working characteristics. Thus, the two types of sheds will vary in looks, ease of modification, and time required for assembly. For example, metal sheds are usually easier to put up and are fairly standardized in form. The main differences among models are in size, roof style (gable, gambrel, or shed), and features such as door-closing systems and protective finishes. By contrast, erecting a wood shed is a more complex process requiring more building skills. But wood sheds are also relatively simple to customize, since the materials are easily reworked and can be used in conjunction with a wide variety of other building materials. In kit form, the shed styles you'll find are likely to be limited, but most can be modified according to your preferences.

Wood Sheds Versus Metal Sheds

	WOOD	METAL
Materials	Structural lumber (occasionally steel) framing pieces. Exterior siding of plywood, particle board, wafer board, or other sheathing material. Materials vary from dealer to dealer.	Usually galvanized steel with heavy-duty enamel finish. Sometimes aluminum. Look for sheds with A.S.T.M. approval.
Completeness of kit	Varies considerably; few kits supply all materials.	All needed parts are supplied except foundation anchors. Flooring is an option; it is not always available.
Cost	From $150 to $700, depending on the size of the structure and the type and quality of materials provided. Foundation materials are often extra.	Generally $100 to $500, depending on size and special features. Anchoring and site preparation are extra.
Tools and time required to install	Varies according to the skill of the builder and the scope of the project; usually several days. Full complement of carpentry tools, in many cases.	One day's time, or less. Screwdriver, pliers, stepladder, plus tools for site preparation, which vary according to the foundation chosen.
Skills needed to install	Good carpentry skills are very helpful and in some cases essential.	Relatively few skills needed.
Possibility for exterior modification	Dimensions and design features can often be modified to meet individual needs. Customized look is easy to create with finish materials, paint, and trim.	Slight—dimensions and appearance are predetermined. Some manufacturers sell optional window kits.
Possibility for outfitting interior	Interiors are easily outfitted with built-in structures or wall-mounted components.	Limited options; accessories are available from some manufacturers.
Disadvantages of the material	Subject to rot if exterior-grade materials are not used. Any wood that contacts the ground should be pressure-treated.	Subject to rust, so moisture and ventilation are important issues. Subject to denting. Lightweight; subject to damage from strong winds or heavy snow loads.
Permanence	If well constructed, long enough life span to be considered permanent.	Relatively short life span (5 to 8 years).
Care and maintenance required	Regular interior cleaning. Periodic repainting or restaining. Regular removal of leaves, twigs, and pine needles. Ventilation.	Regular interior cleaning. Periodic washing of exterior and waxing with good paste wax to help preserve finish. Touch-up painting of any scratches to prevent rust. Periodic tightening of screws and adjustment of doors. Regular removal of leaves, twigs, and pine needles. Removal of snow and ice from roof. Ventilation for heat and moisture.

How Much Work Will It Take?

Before you buy, get a clear idea of how big a project you can reasonably tackle in terms of time and effort. Many home centers offer installation services, a worthwhile alternative if you lack the time, skills, or interest to do the assembly yourself.

Some shed kits supply complete materials and are easy to assemble; others require much more work. The kit should include detailed instructions. Ask to see the booklet before you make the purchase. If it seems incomprehensible to you, choose a different shed.

No matter what type of kit you select, you'll have to do some site preparation and lay a foundation. The site should be a level, firm, and well-drained spot, so the shed will sit squarely, settle evenly, and be subject to minimal moisture.

Depending on your local build-ing code, you have several choices for a foundation: a layer of patio blocks, a bed of gravel, a concrete slab, a pier foundation, or pressure-treated lumber skids. Whichever base you choose, you'll want the shed to be securely anchored to it. Some prefabricated sheds come with flooring, but most do not, so the base you put down will be your floor unless you opt to construct another one.

A Comparison of Project Kits

Metal Sheds	Metal sheds offer the most complete prefabricated kits. Few tools are needed, and the basic skills required are patience and ability to follow the instructions provided. (See pages 62–63.)		
	Package Includes	**Other Materials Needed**	**Comments**
Metal Shed Kit	Instructions. All structural parts and hardware.	Anchors. Flooring, if desired.	Probably the simplest type to build since you don't have to measure or cut materials, but the most limited in terms of design and customization possibilities.

Wood Sheds	Kits for wood sheds can take many forms. Lumberyards and home centers commonly put together their own packages for wood sheds, using materials they have in stock. Parts included can vary greatly, but, in any case, assembly is likely to be a major construction project. Check what kits are available locally. Some of the approaches you might encounter are listed below. (See pages 56–59.)		
	Package Includes	**Other Materials Needed**	**Comments**
Precut Wood Shed Kits	Instructions. Precut framing. Precut sheathing. Hardware. Trim.	Paint; incidental finish materials.	Few tools are required. Assembly is simplified; you don't have to determine, measure, or cut materials.
Partially Precut Wood Shed Kits Note that some types are supplied with metal framing components.	Instructions. Precut wood framing. Uncut sheathing. Hardware. Precut trim.	Finish materials.	Precut pieces may be numbered to make assembly easier but you'll still need to do some measuring. Sheathing materials will need to be cut to fit.
	Instructions. Steel framing components.	Hardware. Sheathing materials. Trim and finish materials.	Steel framing components work like precut wood studs. Some dealers may sell the frame kit with preassembled sheathing and trim materials. Some measuring and cutting will have been done.
Uncut Wood Shed Kits Kits listed here may be supplied with some, or none of the wood required.	Instructions. Cardboard templates. Uncut framing and sheathing materials. Hardware.	Trim and finish materials.	Cardboard templates work like a dress pattern to guide measuring and cutting. Templates may be sold separately, with instructions and a lumber list, so you can choose your own materials. Measuring and cutting are simplified.
	Instructions. Metal connectors for framing. Paper templates.	Framing lumber. Sheathing materials. Carriage bolts and other hardware. Trim and finish materials.	This kit uses simple construction to create a very sturdy shed with a five-sided base. Templates simplify measuring and cutting. Connectors simplify assembly. Dimensions, styling, and materials can be varied to suit your needs.
	Instructions. Uncut framing and sheathing materials. Hardware.	Trim and finish materials.	This approach is the closest to building from scratch with traditional techniques, though you don't have to determine how much material you need.

Wood Shed Kits

Conventional Construction

Lumberyards and home-improvement dealers offer a rich variety of wood shed kits (see page 55). Explore the shed kit options available locally to find out the specific skills you'll need to install one. For some kits, you'll embark on a full-scale construction project, requiring an array of carpentry tools and the ability to use them well. Other precut wood shed kits are as complete as their metal cousins and about as simple to assemble. In general, however, a good deal less preassembly work has been done for you in wood kits.

When you build using wood materials, you can give the shed your personal stamp. While the size and shape of the building are predetermined by the shed's structural design, it's not hard to make functional and aesthetic changes, such as adding a window or moving the door.

On the inside, shelves, hooks, hangers, and work surfaces can be easily fastened to the walls, and you have great latitude to tailor the fittings to accommodate particular uses—by either building in fixtures or buying components (see the preceding two chapters).

Outside, it is comparatively easy to make a wood shed compatible with its setting. Since you choose your own finish materials—paint, stain, moldings, shingles, trim—you have control over the color, styling, and overall look of the shed. Shingled surfaces, for instance, give the shed a different appearance from a painted plywood exterior. Landscaping also helps to integrate the shed visually into its surroundings.

The Building Process

Once you've selected the shed you want, figure out what additional materials you need to buy, if any. When you have everything you'll need for the job, you're ready to start building: lay the foundation and floor; frame the walls, the door and window openings, and the roof; sheath the structure to make it solid and weather tight; apply the roofing material; add any trim; and, finally, finish the shed with stain or paint. To get a more detailed view of the construction process, refer to the next chapter. There, you'll see how to build a wood shed from scratch. The basic steps are the same when building from a kit, but some of the work will have already been done for you.

This shed's gable roof, vertical siding, and window were part of the kit's architectural features. The door is too narrow for a lawn tractor, but the size of the shed and the light from its window make it a good multiple-use space. It can store garden supplies and provide places for potting—indoors on dreary days and outside on the workbench when the weather is fine.

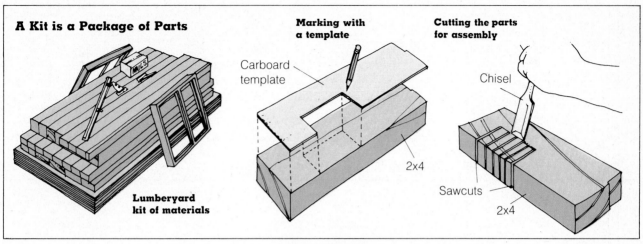

A Kit is a Package of Parts

Lumberyard
kit of materials

**Marking with
a template**

Carboard
template

2x4

**Cutting the parts
for assembly**

Chisel

Sawcuts

2x4

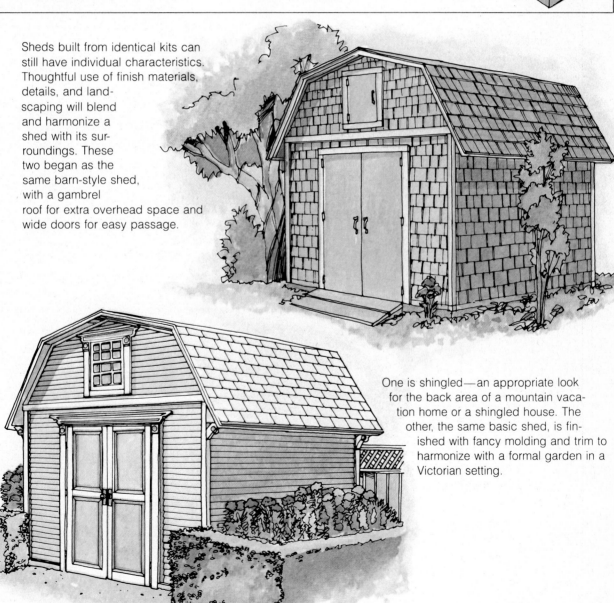

Sheds built from identical kits can still have individual characteristics. Thoughtful use of finish materials, details, and landscaping will blend and harmonize a shed with its surroundings. These two began as the same barn-style shed, with a gambrel roof for extra overhead space and wide doors for easy passage.

One is shingled—an appropriate look for the back area of a mountain vacation home or a shingled house. The other, the same basic shed, is finished with fancy molding and trim to harmonize with a formal garden in a Victorian setting.

Wood Shed Kits

Connector Systems

Shed structures may be framed by bolting lumber struts to special metal connectors—a construction approach resembling a grown-up version of Tinker Toys. Metal connectors are designed to join lumber framing members together into stronger constructions than can be achieved using standard joining techniques.

Two types of building-system connectors are available. One gives you a shelter with triangular walls, much like a geodesic dome. The other yields a vertical-walled structure. It has a more traditional appearance but its graceful five-sided shape distinguishes it from rectilinear sheds.

In either case, this fast and flexible construction system allows you to tailor the size of your building to your spatial needs, working within the confines of the pentagonal base. Your kit will supply an appropriate number of connectors, plans for a variety of structures that can be built from the same kit, and paper templates to guide your cutting and drilling for each one. The first step is planning what type of structure you want for your particular needs, how big it should be, and what materials you'll need. By varying exterior finish details, you can give the same structure quite different looks.

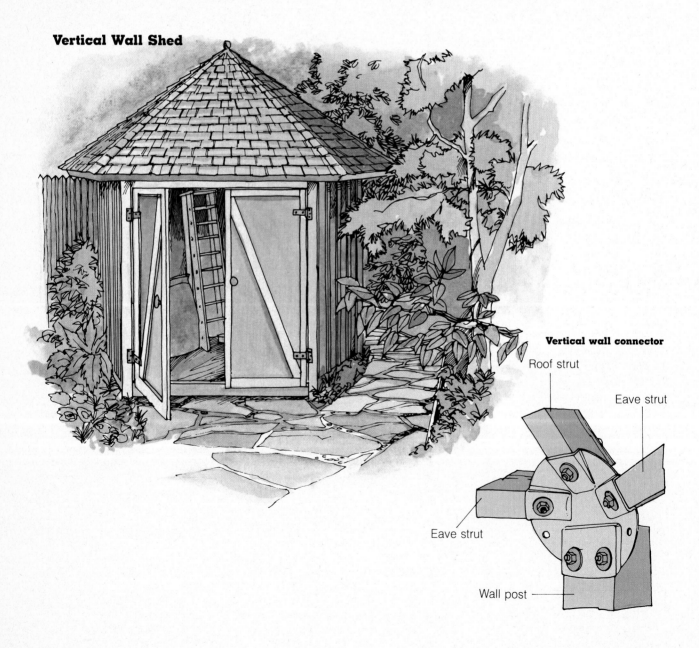

Vertical Wall Shed

Vertical wall connector

Roof strut

Eave strut

Eave strut

Wall post

Triangular Wall Framing

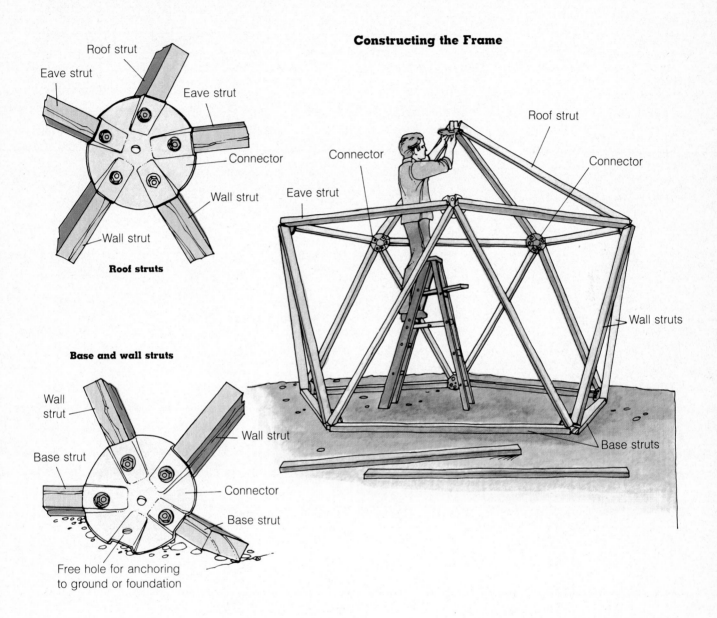

Roof struts

Roof strut
Eave strut
Eave strut
Connector
Wall strut
Wall strut

Base and wall struts

Wall strut
Wall strut
Base strut
Connector
Base strut
Free hole for anchoring to ground or foundation

Constructing the Frame

Roof strut
Connector
Connector
Eave strut
Wall struts
Base struts

The Building Process

Cut struts to length (in the easiest plans, the struts will all be the same length, which simplifies cutting). Then, predrill holes for the bolts as indicated by the templates. As you bolt the struts to the metal connector plates, the structure takes shape.

When the framework is complete, you must anchor it to its base, sheath the walls (or leave them open, if you wish), cover the roof, and construct and hang the door. As the last step, trim and paint or stain the shed.

Because this kit system has you do so much of the planning yourself, you have an opportunity to create a storage building that is attractive and compatible with your yard and house.

A connector-built shed, particularly one with vertical sides, offers the versatility of other wood sheds when it comes to adapting the interior. Sheds with inclined triangular walls don't lend themselves as readily to outfitting with standardized shelves and hangers, but freestanding units work well.

Metal Shed Kits

Prefab Structures

When you think of a metal shed, you probably picture something that resembles the one below—square, sturdy, practical. Metal sheds have the advantage of being inexpensive and easy to assemble. As page 62 shows, metal storage structures can be versatile, too, and several types are available to meet special needs.

Full-size metal sheds typically resemble each other a great deal. Variety comes mainly in color and roof style. A gable roof looks trim, and being a common roof line, echoes the lines of a good many homes. A gambrel roof is a familiar design for barns and country buildings. The roof structure offers extra storage space and may even feature an "attic"—a rack around the tops of the walls where smaller items can be stored.

Vertically ribbed wall panels are characteristic of metal sheds—less for aesthetic reasons than for structural ones. The ribs lend the sheet material the extra rigidity it needs in order to be strong enough for constructional applications.

Look for sheds that meet the voluntary standards regarding structural strength and durability that have been established by the American Society for Testing and Materials.

Options for modifying a metal shed's exterior are limited, so choose a model whose appearance appeals to you, and harmonizes with your home's architectural details. One easy way to blend it into its environment is through careful landscaping. Plantings around the shed will help tie it visually to its surroundings. A second option is to screen the shed with a fence of boards, grapestakes, or lattice.

A metal shed tends to stand out more prominently in the environment simply because the material isn't common to landscape gardens. By taking care to integrate it purposefully and thoughtfully into the surroundings, with pathways, plantings, and other landscape elements, a metal shed can share your outdoor living spaces discreetly and serve your storage needs well.

The Building Process

Metal sheds come in complete packages; assembly demands little knowledge of construction and few tools—a screwdriver, a pair of pliers, a stepladder, an awl for aligning holes, work gloves, plus whatever tools you need to make the foundation and to anchor the structure.

The kit you bring home will contain sheet metal panels, long sections of framing, trim pieces, and lots of small screws, bolts, washers, and other bits of hardware. The first step in transforming them into a shed is to give the structure shape. You'll begin by assembling and connecting the floor frame, the roof frame, and the corner panels. Then bolt the wall panels in place. Next, add the gables, roof, and door. The last step is to anchor the shed either to the base you have prepared for it or directly to the ground. Shed manufacturers sell anchoring kits as options. There are two types: one screws the frame of the building to your lawn; the other uses cable that you string over the shed's beams and then fasten to the ground with anchors.

Typical Metal Shed Assembly

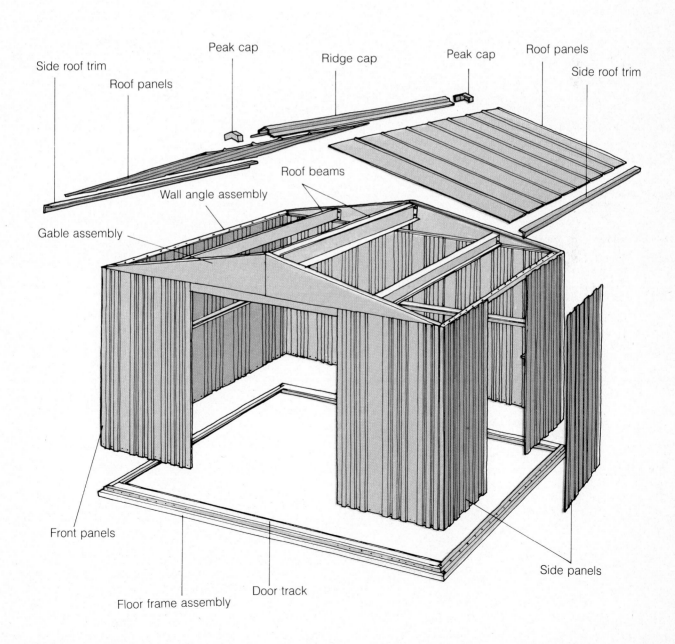

Side roof trim · Roof panels · Peak cap · Ridge cap · Peak cap · Roof panels · Side roof trim · Roof beams · Wall angle assembly · Gable assembly · Front panels · Door track · Floor frame assembly · Side panels

Metal Shed Kits

Prefab Structures

Balcony Storage Box

This heavy-duty steel box can earn its keep by organizing items that are awkward to store, such as firewood, trash cans, fertilizer bags, or outdoor toys. Measuring just 4 feet wide by 2 feet deep, it's a practical idea for a balcony, deck, or other tight spot where you need storage space but have limited square footage to devote to it. The top opens, as well as the double doors on the front, making access convenient. Some models have built-in locks for safe storage of chemicals or sharp tools.

Mini-Shed

Bike Locker

If you have no garage, where can you stow your bicycle to keep it out of the weather and safe from theft? A bike locker is one answer; it will keep two bicycles dry and protected. Small motorbikes can fit in, too—or any items you want to store securely. Besides having lockable doors, the unit is constructed with screwheads and ground anchors inside for extra protection.

For a small yard, choose an appropriately scaled shed. This 4-foot by 5-foot structure, nestled between fence and sidewalk, is an economical option, since the cost of a metal shed is related directly to its size. Its door slides open on an outside track, preserving all of the interior wall space for storage use.

Accessories

While it's tricky to mount things to the framing or wall surfaces of metal sheds, there are two simple ways to increase the shed's storage capacity. One is to outfit it with freestanding units. See the preceding two chapters for ideas. The other way is to buy accessories sold by the shed manufacturer. These might include shelf units, overhead storage racks, hanger bars with hooks, or corner workbenches, all designed specifically for installation in metal sheds. Not all manufacturers make all types, and one maker's products may not fit another's sheds. Check on their availability when you buy your kit.

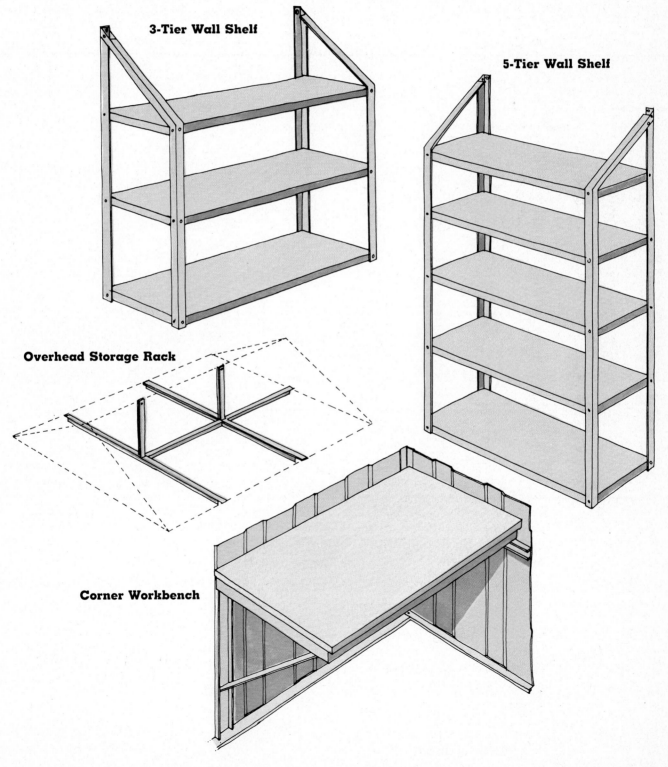

3-Tier Wall Shelf

5-Tier Wall Shelf

Overhead Storage Rack

Corner Workbench

Shed Plans: Structures You Can Build

Building a shed from scratch can be a big—but rewarding—project. This chapter gives plans for a versatile basic shed, shows ways to tailor it to your needs and taste, and tells how to build three other storage structures that might suit your situation.

Building a shed from scratch requires both effort and time; but it yields a double satisfaction. You'll give shape and substance to a key part of the storage plan you've developed and you'll gain the benefits of systematic organization, especially suited to the patterns of your life. In addition, you may enjoy the construction process itself—the feel of the wood, the heft of the tools, and the pleasure of making your project evolve from an idea into a useful, good-looking structure.

You might choose to design the structure or adapt a set of plans to meet your requirements. If you don't want to wield the hammer yourself, but you still want a well-crafted, custom-built shed, you can hire someone to do the construction.

This chapter gives you a detailed plan for a basic shed, then shows you ways you can vary elements of the basic shed to achieve striking differences in the way it looks and functions. You can change the roof style or the type of foundation, for example. You can alter the style and placement of doors and windows (which in turn will affect the way the shed is used—see pages 84–85). You can add an interior wall. Essentially, this basic shed offers you several plans in one—adaptable for a variety of purposes or for several purposes at once.

While a prefabricated shed may be suitable for

storage only, the shed you plan and build can serve more than one function—a storage structure plus a pool cabana, or an outdoor kitchen, or a potting shed, to name only three possibilities that are detailed later in this chapter. You can apply these same construction principles to other building plans you might acquire.

When you build your own storage structure, it need not look like a typical shed—a miniature cottage or a baby barn. Many other solutions exist that will blend storage subtly into your outdoor living area. Plans for three of these structures are included in this chapter:

■ A slender lean-to that adjoins the main house and becomes part of its architecture.

■ A storage wall that divides outdoor areas attractively while it stores the gear used in those areas.

■ A project that takes advantage of the space under an existing roof—in this example, a carport.

If you look attentively around your yard, you will probably spot other places where storage might be accommodated in a way that integrates the new structure into your environment, enhances the area, and gives you the storage space you need.

The projects in this chapter are not necessarily difficult to build, but they do require a commitment of time and effort. You'll need some carpentry skills and an understanding of elementary construction procedures. Ortho's book, *Basic Carpentry Techniques* is a good reference if you want to learn more. It tells about tools, construction materials, and how to use them, and it gives a step-by-step guide for constructing a small building, from laying the foundation to raising the roof.

◀ This neat wood shed captures the rustic mood of its almost-country location. Its foliage-green color fits well in the woodsy site. Sheds designed and built from scratch, as this one was, are all comparable in their esssential form and construction. The basic shed plan, variations, and other suggestions in this chapter can help you erect a structure that works as well in your location as this one does here.

Before You Build

Select a Shed Plan

This chapter includes plans for constructing a basic storage shed, the construction steps you'll need to take, and ideas on ways you can vary it in several ways to suit your needs. In addition, the chapter guides you through the construction process for three smaller storage structures.

There are, however, many other shed plans put out by publishers, building material companies, trade associations, and do-it-yourself magazines. On page 94 you'll find a list of outdoor storage resources from which other shed plans can be purchased.

Keep in mind the following factors as you determine which plan will be the right one for you:

■ The extent of your own construction skills.

■ The amount of time and money you want to spend.

■ How you want your structure to function.

■ How you want it to look.

■ How you want it to relate to its site and the rest of the yard.

■ The structural requirements of local building codes and climatic conditions.

If you don't find the perfect plan, don't worry. Differences between one shed and the next are mainly matters of size, shape, and detailing; modifications aren't hard to make. If you've had experience in building and want to put some time into the planning, you can design a simple structure on your own. If you lack that experience, but want to achieve a specific functional solution, or create a certain architectural look, consider hiring an architect, a building designer, or a contractor to draw up a set of plans.

What Are Your Resources?

Think about the amount of time, money, and effort you want to expend on your project. Building your own shed is going to cost you some of each.

The small-scale projects in this chapter could be erected in one weekend. They're relatively simple to build and are good choices if your skills are still developing. Because they use fewer materials, they cost less. The full-size shed requires greater skill and a more substantial investment all around.

What Are the Shed's Functions?

Whether you design your own structure, adapt an existing plan, or follow a plan in this chapter, you want the new structure to be part of your overall storage system. At the same time, consider how the building might contribute to the smooth functioning and enjoyment of activities in your yard. For example, is it to be strictly a storage shed? Or would it be more practical if it had multiple uses? Perhaps it could incorporate a workshop, a hobby area, or a swimming pool dressing room. If it's to house your barbecue grill and equipment, you might want to create an outdoor kitchen next to the patio for convenient entertaining. The process outlined in the first chapter will help you define your needs, develop a storage system, and determine what functions the new shed will serve.

The purpose of the structure prescribes how big it should be and what features it should contain, so have its functions firmly in mind when you choose a plan. It is usually much simpler to build a particular feature into your shed in the first place than to add it later.

As you plan your shed, give some thought to the architectural features you want it to have:

■ **Interior walls** to divide the space for different uses (see page 84).

■ **Size and placement of doors and windows.** See page 84 for some ideas about how these affect the use of the space.

■ **Electricity or plumbing** or both might make the shed more functional for a number of

Check Building Codes

Building codes regulate how far from the lot line your shed must be, how tall it can be, and what percentage of your total lot can be covered with construction. Believe it or not, building codes and permits weren't created just to be one more bureaucratic annoyance. Their main purpose is to ensure that any alteration or addition to your home is safe for use. Your local building codes can tell you how to make your building structurally sound, so that it will stand up to the rigors of your climate, and they help you reduce the risk of fire.

Inquire about permits before you begin building. Requirements vary considerably from place to place. In some localities you don't need a permit for a shed under 120 square feet, but in others you need one for any sort of construction. To obtain a permit, you'll have to submit copies of your plans to the building inspector for approval.

In some neighborhoods, homeowners' associations must review plans for any changes to the exterior of a home, to guarantee that the alteration is in keeping with community standards. If you happen to live in a designated historic district, you may also need to comply with special requirements designed to preserve the area's unique character.

purposes. For guidance, see Ortho's books *Outdoor Lighting, Basic Wiring Techniques,* and *Basic Plumbing Techniques.*

■ **Insulation** for a more comfortable work area, or to protect sensitive possessions from extremes of heat and cold. See Ortho's book *Energy-Saving Projects for the Home.*

How Do You Want It to Look?

When you build from scratch, you have the chance to tailor the appearance of the structure to your taste and preferences. While your shed should be serviceable, functional, and well-crafted, it need not be a focal point in your yard. It can blend quietly into the setting with a pleasing appearance. One good way to achieve this aim is by using the same architectural details on the shed that appear on your house—rooflines, trim details, window styles, exterior finish materials, and the color scheme all suggest sound ideas for creating a good-looking site addition.

There are no hard-and-fast rules for design—it can be a creative process for you just as it is for an architect. Throughout this chapter, you'll find ideas to get you started. To help you visualize what you want, make some sketches of the shed, playing with various rooflines, finish materials, window styles, and color schemes, until you come up with a look you particularly like.

Some smaller storage structures blend so successfully with their surroundings that they become practically invisible—no one identifies them as outdoor storage elements. They may be tied in with an existing feature in the landscape or given a design function of their own, while their warehouse role is played down. See the storage wall on page 90. Storage needs are met, yet that function is whispered, not shouted.

Choose the Right Location

In the right spot, a shed is a beneficial addition to your yard. In the wrong place, it can become an additional structure on your property that you aren't able to utilize fully, and in that event, a hindrance to your storage system. Here are some factors to consider when selecting the location.

Convenience and Access

Since you want the shed to be handy, pick a location convenient to areas where the stored items will be used—near the garden if it will house garden tools, near the play yard if it will contain the kids' outdoor toys. Consider whether the things to be stored can be taken out and put away easily. For example, you don't want to have to turn sharp corners to maneuver a lawn tractor in. Think about access in bad weather, too. If you must use the shed regularly despite rain and snow, put it close to the house so you won't get drenched each time.

Traffic Paths and Activity Areas

A shed will affect how you use the rest of your yard. What routes are people likely to take when they go to the shed from various parts of the house and grounds? Will you be creating traffic paths where you don't want them? Will a shed in that spot displace other important activities?

Sun and Shadow

The shed's orientation to the sun will affect how much heat and light it takes in. Having a long side facing south gives the most moderate interior temperatures. Think also about where shadows from the shed will fall. For example, you don't want the shed to block the sun from your tomatoes or rosebed.

Appearance and Views

Consider how the shed's location will contribute to the appearance you want for your yard. Will the shed dominate the landscape or be a minor feature? Will you see the shed instead of the apple tree whenever you peer through your dining room window, and does that matter? How about the view your neighbor will have—will it still be pleasant?

Terrain

Choose a spot that is flat, firm, and dry so that the building will sit squarely, settle evenly, and drain naturally. Be sure that you've allowed enough room to move about as you're doing the construction.

Putting up your shed will be much simpler if it is on even ground. While a post-and-pier foundation and some other types can be used on a hilly site, building on a slope is more difficult; you may need to hire professionals to help.

Avoid soft ground. In addition to causing excessive or uneven settling, it may indicate that the site, is too damp to ensure a long life for your shed.

Moisture can be harmful to both your shed and its contents, so the site should be a well-drained spot that doesn't collect standing water. If necessary, grade it so that any runoff will flow away from the building.

Before You Build

Consider the Climate

The appropriate structural design for your shed depends a lot on the climate where you live. The structure has to be able to withstand the onslaught of harsh weather. One purpose of building codes is to set out proper construction practices for the locality, given its amount of snowfall, its susceptibility to hurricanes and tornadoes, and other factors of climate.

Rain and Moisture
Moisture, from rain and other sources, can rot the wood, rust the metal, and spoil the contents of a shed. Your aim is to keep wetness out by preventing flooding, seepage, and leaks.

Dampness is most likely to creep in at the top and bottom of the building. If your foundation is a concrete slab, a sheet of 6-mil plastic between it and the earth will serve as a vapor barrier to prevent moisture from percolating up into the building. The roof should slope enough for rain to run off it easily. A shingled roof needs a greater slope than one covered with sheet material, since each joint or break in the surface continuity is a place where water can work its way in.

A wide roof overhang will carry the runoff farther from the building, protecting the foundation and the surrounding earth from erosion or puddles of standing water. Unless the wind is strong, the overhang will also inhibit rain from hitting the walls directly, reducing the effects of weathering. Flashing, caulking, sealants, and other materials used in the construction of your shed help to make it water-tight.

Snow and Freezing
Snow affects the stability of a shed from above; freezing affects it from below. The principal danger from snow is its weight. The

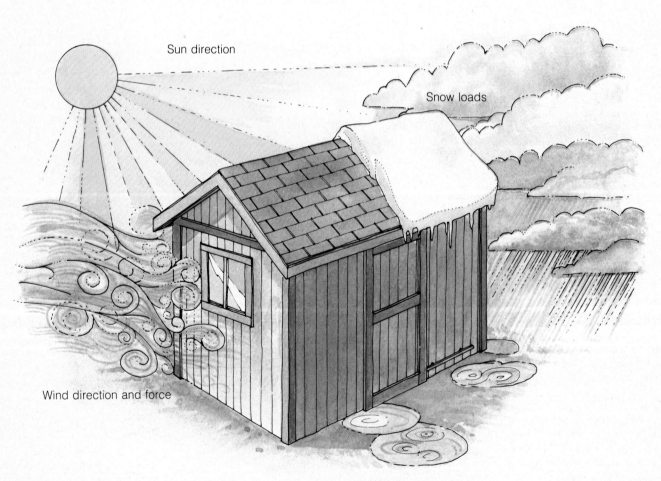

Sun direction

Snow loads

Wind direction and force

Water and moisture

roof must be strong enough to bear the maximum load. To achieve this, rafters must be properly sized and spaced. Usually they are made of 2 by 4 or 2 by 6 lumber, set no more than 2 feet on center. A steeply pitched roof sheds snow more easily than a flatter one. Your local code will tell you what's required in your area.

Freezing causes the ground to heave, which can make the building skew on its foundation and pull apart. Foundation footings should sit on solid earth, below the frost line. Surface foundations, such as wood skids or piers without footings, aren't subject to frost heave, since they rest *on* the earth, not in it.

Heat and Humidity
Heat buildup can make a small, enclosed space, such as a shed, very uncomfortable. Add high summer humidity and you have an excellent breeding ground for mildew and other fungi. Proper ventilation will help relieve both problems. At a minimum, you'll want vents in the gables, even if your shed has a window. Additional vents near the foundation help by enabling crosscurrents of air to circulate through the shed.

Wind
Wind exerts two kinds of forces on a structure: lateral loading against its walls, and uplift—raising it from the ground. Walls that are properly braced will resist the lateral force. Plywood sheathing provides excellent bracing because of its inherent diagonal (shear) strength (see page 70). Walls surfaced with other materials may need added diagonal bracing. The answer to uplift is to anchor the structure securely to the earth. Building code standards for bracing and anchoring are determined by the speed and direction of prevailing winds in your area and by the likelihood of severe windstorms.

Prepare the Site

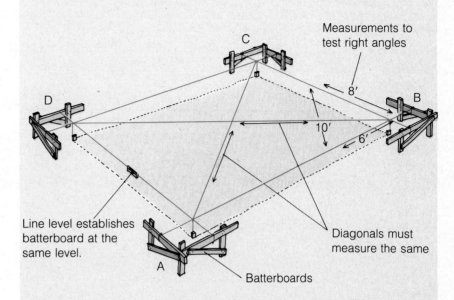

Measurements to test right angles

8'

10'

6'

Line level establishes batterboard at the same level.

Diagonals must measure the same

Batterboards

C

D

B

A

Excavating and Grading
If you've chosen a flat, well-drained site, preparation is a simple process of leveling and smoothing the soil and clearing out roots and rocks.

The key word is *level*—the points where the building bears against the ground must be perfectly level. With a concrete slab, that means the whole foundation. With piers or skids, it means spots where parts of the structure touch the ground; in other places the soil needn't be smooth. Piers and skids must also be set so that their top surfaces are all in the same level plane, so that the shed will rest squarely on them.

If you have to do any amount of digging to prepare the site, check first with local authorities on the location of underground pipes and gas and electric lines.

Staking Out the Building
Establish the corners and boundaries of your shed, and make certain the layout is squared. When the diagonals between pairs of opposite corners are equal in length, you'll know that this has been achieved. Check that the shed layout is also properly aligned with the house and with fences or property lines.

Making the Foundation
Three types of foundations are shown in this chapter: wood skids, a concrete slab, and a pier foundation. While they have different features and advantages (see pages 78–79), the purposes of each are the same: to hold the building level and square, to bear its weight on the soil, to help it settle evenly, and to protect it against moisture, freezing, and other problems.

Lumber

Most lumber used in construction is softwood—wood that comes from coniferous trees such as cedar, pine, spruce, larch, Douglas fir, or redwood. Lumberyards in different areas of the country tend to stock heavily the wood species grown in that region. Other species are often available, though some may have to be ordered specially.

Softwood lumber falls into two major grading categories. *Select lumber* is graded for appearance. For construction, you want *common lumber*, which is graded for strength.

There are four categories of common lumber:

■ **Select structural lumber** receives top marks for both appearance and strength. It is divided into five grade levels; number one grade is the highest quality.

■ **Structural joists and planks** follow the same numerical grading system as select structural lumber.

■ **Light framing lumber,** the type most widely used for framing houses, has three grades. Construction grade is the strongest and has the fewest knots. Standard grade comes close to construction grade in quality but is less expensive. Utility grade is likely to be weak

and knotty; it isn't up to the structural requirements of most building codes.

■ **Stud grade lumber** is suitable for vertical framing, although it shouldn't be used for horizontal framing members. It's sold precut to standard stud lengths.

Lumber is sold according to *nominal* dimensions—for example, 1 by 6, 2 by 4—that tell you the size of the piece when it was first cut, before it was dressed (surfaced). The *actual* dimensions of the board are somewhat smaller. Thus, a 1 by 6 really measures 3/4 inch by 5 1/2 inches, while a 2 by 4 measures 1 1/2 inches by 3 1/2 inches. Lumber used for framing is generally 2 by 4 or 2 by 6, and 4 by 4s are commonly used for posts.

When buying lumber, avoid green wood with a moisture content higher than 29 percent; it's likely to warp as it dries. Steer clear also of defective boards—ones that are crooked, twisted, split, or full of knots.

Pressure-Treated Wood
Pressure-treated wood resists damage from decay and insects, making it an appropriate choice for structural members that directly contact the earth or come close to touching it. Chemical preservatives are forced under pressure into the wood's cells, enabling treated lumber to last many years longer than un-

treated wood, even resistant species. Many building codes require pressure-treated wood for certain applications.

Treated wood that is to go directly in or on the soil contains more preservative than wood intended for use above ground. Most treated lumber bears a quality-control stamp with an alphanumeric code. A two-digit number (such as LP–22) indicates that the lumber is suitable for ground contact, while a single digit (LP–2) means above-ground use only.

Pressure treatment doesn't change the wood's basic characteristics. It will still weather in the same way and look the same when painted or stained. It is, however, chemically toxic, and precautionary handling advice should be heeded.

Plywood

Plywood is a good choice for shed flooring, siding, and roof sheathing. It's durable, and because it comes in large sheets—usually 4 feet by 8 feet—it covers big areas quickly. Most of all, it's strong. Thin sheets, or plies, are glued together face to face with their grains running in alternating directions. This makes plywood resist warpage and gives it great diagonal (shear) strength, which yields good bracing.

Plywood is commonly labeled with two letters, ranging from A to D, which indicate the quality of its two sides. A is the highest grade; it has virtually no knots or defects. D is the lowest. The letters are always paired, for example, AC. The first represents the side you would put facing out, the second, the inside surface, which is usually not exposed to view and can therefore be a lower grade. A third letter, X, means the sheet is suitable for exterior use. Exterior-grade plywood is manufactured with moisture-proof glue; often its outer plies are made from decay-resistant woods.

Plywood comes in many thicknesses. Most common are 3/8-inch, 1/2-inch, 5/8-inch, and 3/4-inch. The 3/8-inch panels are easy to handle, least expensive, and quite adequate for siding uses. For roofing, a 1/2-inch sheet has greater structural strength and won't sag under gravity or snow loads. For floors, you need 5/8-inch or 3/4-inch plywood, depending on the expected loads.

Concrete and Piers

If you intend to pour a concrete slab foundation, have the concrete delivered by a ready-mix truck.

For smaller quantities of concrete, such as you'd need for footings under piers, it is practical to mix your own, either in a rented mixer (which speeds the work) or by hand in a wheelbarrow. Mix it in the proportion of one shovelful of portland cement to two shovelfuls of sand and three of gravel. Gradually add water (approximately five gallons per sack of cement), and blend it with a mortar hoe until the mixture is the consistency of a thick milkshake, neither watery nor crumbly or stiff.

A pier is a cast concrete unit into which a block of decay-resistant wood was set when the pier was manufactured. Piers form the bearing points for standard floor framing. Precast piers can be bought from your lumberyard. If the ground is uneven or if you need to raise the structure higher than the height of the pier, you can toenail a post of the necessary height to each wood block to make a post-and-pier foundation.

Hardware and Incidentals

You are likely to use four types of hardware in building your shed: fastening hardware, functioning hardware, finishing hardware, and flashing.

Fastening hardware includes the actual fasteners—nails, lag screws, anchor bolts—as well as connectors that facilitate the joining of two pieces of lumber, such as joist hangers, post caps, T-straps, and corner braces.

Functioning hardware includes hinges, door slides, and door closers, while finishing hardware includes doorknobs, locks, and any decorative pieces. All metal parts for the outside of the building should be made of non-corrosive material or be coated to prevent rust.

Flashing is used to prevent water from working its way into the building where different materials meet, or where different parts of the structure come together—along the roof ridges and valleys, around vents, doors, and windows, and at points where walls meet the roof. If your shed is to be a lean-to against the house, you'll need flashing where the new walls join the existing ones, for example. Flashing is often made of metal, but it can also be plastic or asphalt-saturated building paper, as in the case of roofing felt. Galvanized steel and terne-

plate (steel coated with an alloy of lead and tin) are the most common metal types. The alkalies in mortar will corrode galvanized steel, so it shouldn't be used against a masonry wall.

Plan to purchase the incidental materials you'll need, such as caulking, glue, primer, shim stock, and weatherstripping.

The Basic Shed

This shed is a simple, ample, and versatile structure, which can be used for storage alone or for storage combined with work or play activities.

An experienced builder with a helper could put the shed up in a day or day and a half. A novice without help should plan on at least six days—either a week straight or three consecutive weekends.

To simplify construction using common building materials, the shed's dimensions are 12 feet by 8 feet. To alter the size easily, change the dimensions in 4-foot increments.

Selected design options are given on pages 78–83: concrete slab and pier foundations, a gambrel roof and a classic shed roof, alternative types of doors and windows. Pages 84–85 show how the basic shed can be modified to function in different ways depending on how its doors, windows, and interior spaces are arranged. Detailed construction plans for the shed shown here are given on pages 72–77.

Materials List—The Basic Shed

The materials needed are listed below, arranged in order of construction steps. The list gives the quantity and size of each item to buy and a description of the item that includes any special instructions. Try to have all the materials on hand before you start to build.

Part	Quantity	Size	Description
Floor	2	6x8 — 12'	Pressure-treated skids
	10	2x8 — 8'	Floor joists
	2	2x8 — 12'	Rim joists
	3	4x8 — ⅝"	Tongue-and-groove CDX plywood flooring
Walls	54	2x4 — 8'	Cut 34 pieces to 7'2" for studs; use remainder for end plates, sills, trimmers, gable studs, blocking
	6	2x4 — 12'	Wall plates
	1	4x6 — 12'	Header stock
Roof	18	2x6 — 7'	Rafter stock
	1	1x8 — 14'	Ridge board
	3	2x4 — 8'	Collar ties and blocking
Sheathing	11	4x9 — ⅜"	Plywood siding, rough-sawn with kerf cuts 8" on center (if 4x9 sheets are unavailable, use 13 4x8 sheets; locate joints along the bottom edge of the long walls)
	6	4x8 — ½"	CDX plywood roof sheathing

Part	Quantity	Size	Description
Trim	2	1x8 — 14'	Fascia boards
	14	1x4 — 12'	Trim (corners, windows, door)
	2	tubes	Caulking
	2	tubes	Construction adhesive
Hardware	1	8'	Barn door tracks, with
	1	4'	mounting brackets and guide wheels for doors
	2	12"x12"	Screened louvered vents
	2	3'x3'	Aluminum sliding window
	5 lb	6d	Galvanized box nails
	5 lb	8d	Galvanized common nails
	12 lb	16d	Common or sinker nails
Roofing	1 roll		15 pound asphalt saturated building paper
	6	10'	Metal drip edge
	5 bundles		Composition roof shingles (168 square feet)
	3 lb	¾"	Roofing nails
Paint	2 gal		Paint or stain

Anatomy of the Basic Shed

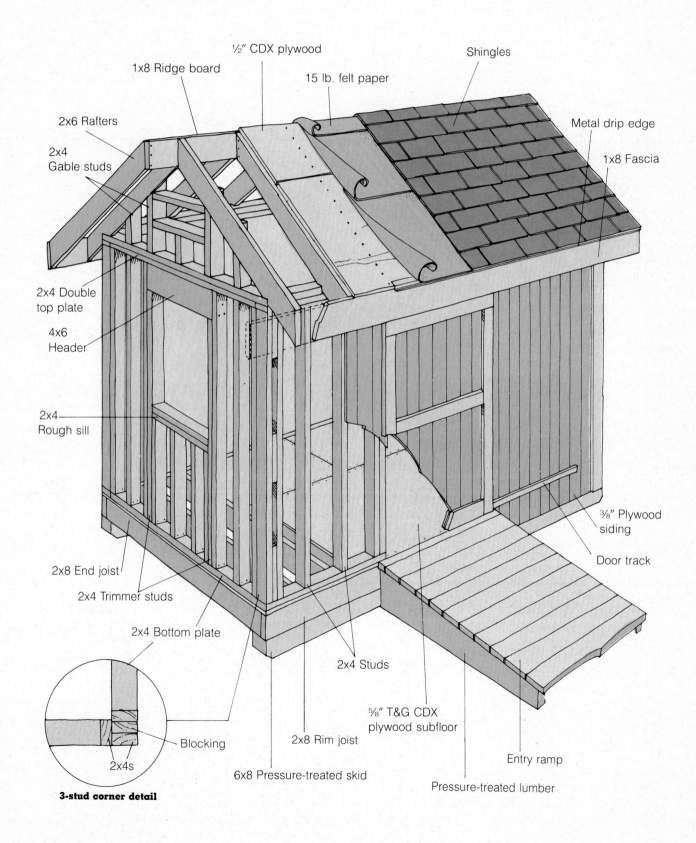

½" CDX plywood

1x8 Ridge board

15 lb. felt paper

Shingles

2x6 Rafters

Metal drip edge

2x4 Gable studs

1x8 Fascia

2x4 Double top plate

4x6 Header

2x4 Rough sill

⅜" Plywood siding

Door track

2x8 End joist

2x4 Trimmer studs

2x4 Bottom plate

Blocking

2x4s

2x4 Studs

2x8 Rim joist

⅝" T&G CDX plywood subfloor

Entry ramp

6x8 Pressure-treated skid

Pressure-treated lumber

3-stud corner detail

The Basic Shed

Foundation and Floor Construction

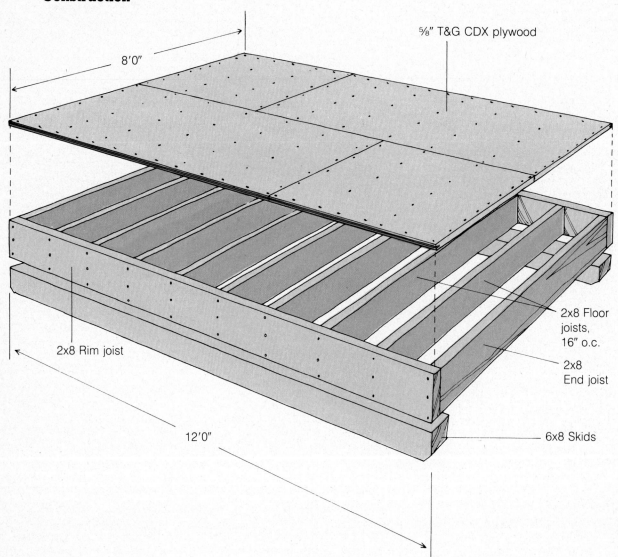

8'0"

⅝" T&G CDX plywood

2x8 Floor joists, 16" o.c.

2x8 End joist

2x8 Rim joist

12'0"

6x8 Skids

Construction Steps

1. Foundation

Place the two 6 by 8 skids on the ground so that they are level, individually and with each other, and their outside edges are 8 feet 0 inches apart. (For better drainage, place each skid over a gravel-filled trench, 12 inches wide by 4 inches deep.)

2. Floor

Nail the 2 by 8 rim joists to both end joists to form a rectangle, using three 16d nails at each joint. Check to see that the assembly is square by measuring both diagonals; they should be equal. Then toenail the end joists to each skid, using three 16d nails at each end.

Nail the remaining floor joists between the rim joists, 16 inches on center, using three 16d nails at each end. Toenail the ends of each floor joist to the skids with three 16d nails; then toenail the rim joists to the skids at 24-inch intervals. Check the platform for level, and make any final adjustments to the skids.

Nail the 5/8-inch plywood flooring to the floor joists, using 8d common nails, 6d common nails, or 6d ring-shank nails, 6 inches on center at the edges and 12 inches on center in the field. Lay the plywood sheets perpendicular to the joists, with a 1/16-inch gap where the sheets butt end to end and a 1/8-inch gap side to side. This allows the material to expand.

Front Wall Construction

12'

7'6½"

Back Wall Construction

Corner framing details

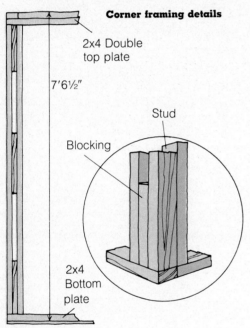

2x4 Double top plate

7'6½"

Blocking

Stud

2x4 Bottom plate

End Wall Construction

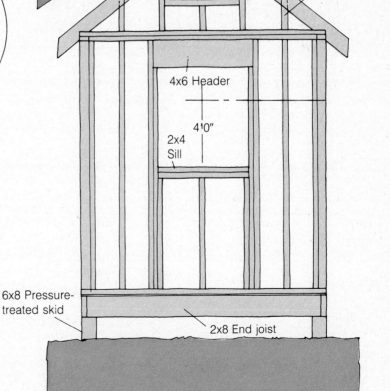

Rafter

Notched 2x4s

12
9

4x6 Header

4'0"

2x4 Sill

6x8 Pressure-treated skid

2x8 End joist

3. Wall Framing

To frame each wall, assemble the studs, bottom plate, and top plate, face-nailing each joint with two 16d nails. Raise the walls into place and temporarily brace them with 1 by or 2 by boards. Nail the bottom plates into the joists below with 16d nails, 16 inches on center. Connect the walls to each other at the corners with 16d nails, 24 inches on center. When all four walls are up, nail a cap plate down onto the top plate, using two 16d nails, 16 inches on center. Be sure the ends are lapped.

The Basic Shed

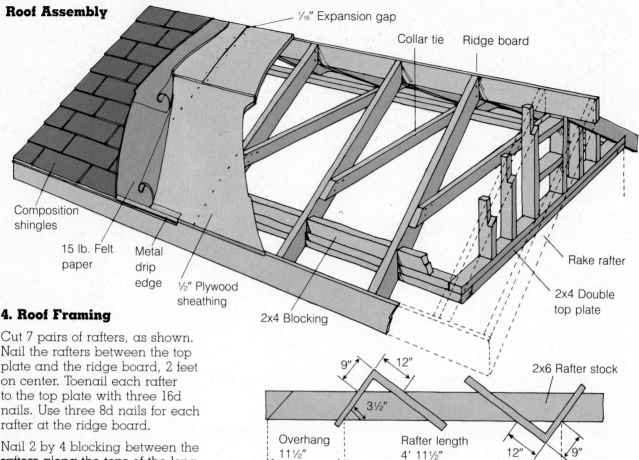

Composition shingles

15 lb. Felt paper

Metal drip edge

½" Plywood sheathing

2x4 Blocking

1/16" Expansion gap

Collar tie

Ridge board

Rake rafter

2x4 Double top plate

9" 12" 3½" Overhang 11½" Rafter length 4' 11½" 12" 9"

2x6 Rafter stock

4. Roof Framing

Cut 7 pairs of rafters, as shown. Nail the rafters between the top plate and the ridge board, 2 feet on center. Toenail each rafter to the top plate with three 16d nails. Use three 8d nails for each rafter at the ridge board.

Nail 2 by 4 blocking between the rafters along the tops of the long walls.

Fasten collar ties to every other pair of rafters using three 16d nails at each end.

Install gable studs, 16 inches on center. Toenail each stud into the plate with four 8d nails, and face-nail the notched end to the rafter with two 16d nails.

5. Window Installation

First staple an 8-inch-wide strip of roofing paper to the bottom of the window opening, making the top of the paper flush with the edge of the opening. Then staple 8-inch-wide strips of paper along the sides of the window opening, lapping the bottom strip and bringing the inside edges flush to the window opening. Nail the window flanges to the framing with 6d galvanized nails or roofing nails, 12 inches on center. Be sure the window is level and plumb. Fi-

nally, staple the top strip of paper in place so that its bottom edge laps over the top flange of the window.

6. Wall Sheathing

Nail plywood siding to the exterior walls with 6d galvanized box nails, 6 inches on center along the edges and 12 inches on center in the field. Cut the door opening flush with the framing. Notch the siding around the rafters.

7. Roof Fascias and Rake Rafters

Nail fascia boards to the ends of the rafters with two 8d nails at each rafter. Bevel the top of each fascia as needed. At each end, overhang the fascia 1 foot past the last rafter.

Cut 2 pairs of rake rafters, as shown in the rafter detail, but

without the bird's mouths. Nail them to the ends of the over-hanging ridge and fascia boards, using three 8d nails at the end of each rafter. Predrill nail holes to prevent the ends of the ridge and fascia boards from splitting.

8. Roof Sheathing and Roofing

Nail 1/2-inch plywood roof sheathing to the rafters with 6d nails, 6 inches on center along the edges and 12 inches on center in the field.

Nail metal drip flashing to the eaves before applying the roofing underlayment. Staple the felt paper horizontally, starting with a 12-inch to 18-inch strip along the eaves for double coverage. Roll out and staple the remaining full-width strips, starting again at the eaves. Nail metal drip flashing over the paper along the rake edges.

Apply composition shingles according to the manufacturer's directions.

9. Vent Installation

Cut a vent opening in each gable end. Caulk around the opening; then nail the flanges of the vent to the plywood siding. Use roofing nails or 6d galvanized nails if the plywood is backed by a stud.

10. Door Installation

Build the door from plywood siding and 1 by 4 trim material. For greater rigidity, both glue and nail the perimeter trim pieces, using an exterior construction adhesive.

Attach the top (8-foot) and bottom (4-foot) door tracks to the shed wall, by bolting the brackets to the studs with 5/16-inch by 2-inch lag bolts or screws. Attach the guide wheels to the top of the door, and hang it. Tighten or loosen the guide wheels to adjust the door up or down and plumb it.

11. Trim and Finish Work

Cut the trim material to size, caulk, and nail the trim pieces on corners and around door and window openings. Build a short ramp for smooth access; pour a concrete ramp, or strap planks together by nailing lengths of 2 by 4 across their undersides. Paint or stain the shed.

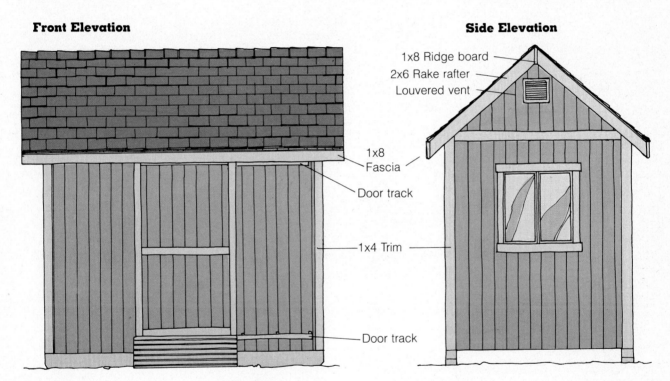

Front Elevation

1x8 Fascia
Door track
1x4 Trim
Door track

Side Elevation

1x8 Ridge board
2x6 Rake rafter
Louvered vent

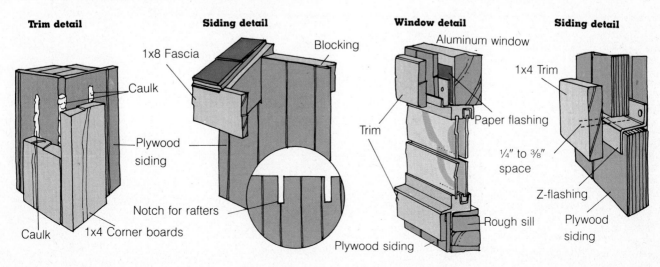

Trim detail

Caulk
Plywood siding
Caulk
1x4 Corner boards

Siding detail

1x8 Fascia
Blocking
Notch for rafters

Window detail

Aluminum window
Trim
Paper flashing
Rough sill
Plywood siding

Siding detail

1x4 Trim
¼" to ⅜" space
Z-flashing
Plywood siding

Alternative Foundations

Piers

Instead of building the shed on skids, construct a pier foundation using precast pier blocks, 4 feet on center. First, lay out the locations of the piers: stretch nylon string lines (representing each wall's outside edge) and secure the strings to batter boards. Mark each pier's location and dig footing holes (a minimum of 6 inches deep and 12 inches square). Fill the holes with concrete, and set pier blocks 1/2 inch deep in the fresh concrete before it sets, making sure they are level and aligned with each other.

The girders themselves will rest directly on the blocks or on short posts, depending on how level the site is and how high off the ground the floor joists must be. Some local codes may require a minimum 18-inch clearance below the floor joists; otherwise, the girders should be at least 8 inches above grade. In the first case, set 4 by 6 girders directly on the pier blocks and toenail them into each block with four 8d nails. In the second case, cut a short post for each pier, and toenail it to the pier with four 8d nails; then connect the girders to the posts with steel connectors. Proceed to construct the shed on the girders the same way as it is built on skids. (Note: See Ortho's book *Basic Carpentry Techniques.*)

Concrete Slab

The shed can be built directly on a 4-inch-thick concrete slab. This technique lowers the shed's height, combines foundation and floor construction into one operation, and can eliminate the need for a ramp.

To prepare the earth bed, remove sod and other organic debris, level the high spots with a flat shovel, and tamp the bed down. Then stake 2 by 8 form boards around the outside edge of the slab. Be sure the forms

Pier Foundation Construction

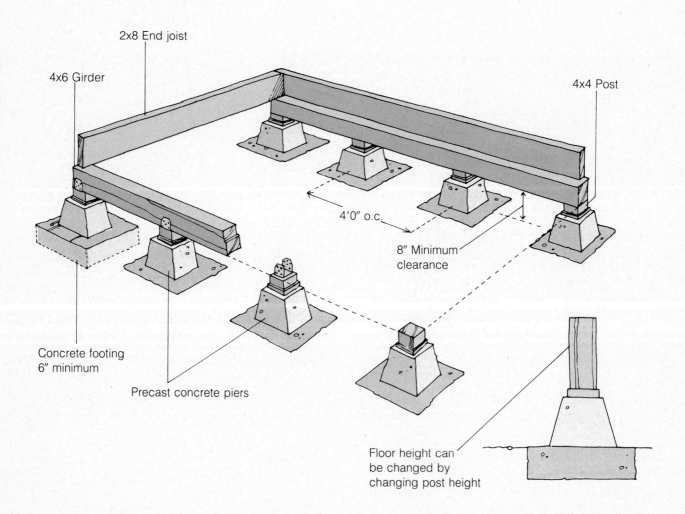

2x8 End joist

4x6 Girder

4x4 Post

4'0" o.c.

8" Minimum clearance

Concrete footing 6" minimum

Precast concrete piers

Floor height can be changed by changing post height

are level and all four corners are square (check by measuring the diagonals). Since the edges of the slab will serve as footings for the shed walls, dig a trench 8 inches wide and 12 inches deep (or to frost line) inside the forms for concrete footings. Then build up the center area with gravel or sand, lay down plastic sheeting to make a waterproof membrane, and cover it with 2 inches of sand (see the illustration). Finally, place steel reinforcing mesh on 2-inch concrete blocks, called *dobies*, or other supports that raise the mesh off the ground, so that it will be sandwiched in the 4-inch

layer of concrete. Now you are ready to pour the slab.

A slab 8 feet by 12 feet, with footings 12 inches below grade, will require 2.25 yards of concrete. For this amount, order a ready-mix delivery, and arrange for extra wheelbarrows or a pump truck if the concrete truck cannot back up to the site.

Once the concrete is in the forms, screed it off by dragging a 10-foot 2 by 4 back and forth with a sawing motion across the tops of the form boards. Then smooth the surface of the slab with a wood float, and place anchor bolts into the concrete, 4 feet on center around the pe-

rimeter, before it sets up. When it has set, apply the final finish with a steel trowel or broom, depending on the desired texture.

Shed materials and construction steps are the same as with a wood frame floor, with the exception of the sole plates. First, they must be a rot-resistant material, such as redwood or pressure-treated lumber. Second, they are held in place with anchor bolts, so holes must be predrilled in the plates. Then, when the walls are raised, to fasten the framed walls, position each plate over the anchor bolts; bolt the wall to the slab.

Concrete Slab Construction

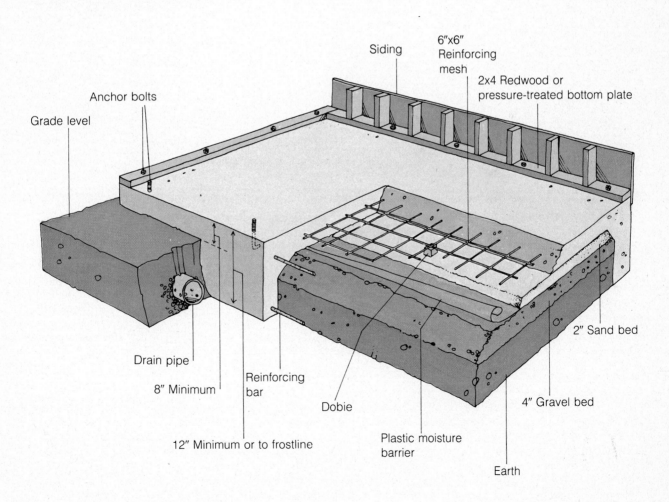

Siding

6"x6" Reinforcing mesh

2x4 Redwood or pressure-treated bottom plate

Anchor bolts

Grade level

Drain pipe

8" Minimum

Reinforcing bar

Dobie

12" Minimum or to frostline

Plastic moisture barrier

Earth

2" Sand bed

4" Gravel bed

Alternative Roofs

Shed Roof

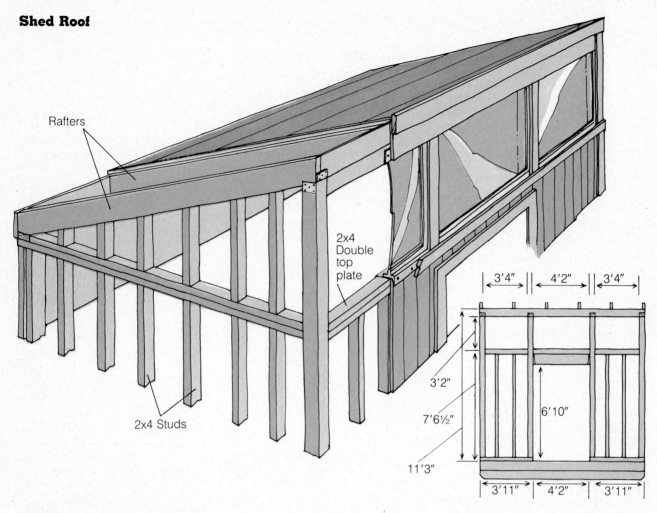

Rafters

2x4 Double top plate

2x4 Studs

3'4" | 4'2" | 3'4"

3'2"

6'10"

7'6½"

11'3"

3'11" | 4'2" | 3'11"

A flat, sloping roof, or "shed" roof, changes the look of the shed dramatically; it also creates a good place for a row of clere-story windows across the top of the tallest wall. With the windows up high and out of the line of visual access, the shed's contents aren't open to view. And unbroken wall space gives extra storage surface area. This entails some changes from the basic shed's construction.

Frame the back (shorter) and side walls in the same way, but make no side wall window openings, and increase the lengths of the filler studs at the top of the wall so that they conform to the new roof shape. The front (taller) wall is framed quite differently, however, using post-and-beam construction to create large window openings across the top. This wall should be built

and raised first, since it is the largest. It consists of a 4 by 6 beam along the top of the wall to support the rafters. This beam rests on four 4 by 4 posts. Place one at each corner and space the other two so that the 4-foot doorway fits between them. Attach a 4 by 6 door header to the posts with metal beam hangers. Check your local code to see if trimmer studs are also required; if so, the spacing between the 4 by 4 posts will have to be wide enough to accommodate the extra trimmer studs. The rest of the wall consists of 2 by 4 studs, nailed 16 inches on center between the bottom plate and double top plates. To frame in the proper size rough openings for your clerestory windows, fill in studs between the top plates and the beam. Since the posts and beam are critical bearing

members, they should be joined with metal connectors at every joint. Raise this wall, and join the others to it in the same way as for the basic shed.

The roof rafters require two bird's-mouth cuts. The easiest way is to cut a pattern rafter; hold it against the ends of the two long walls, with its bottom edge at exactly the same height as the "uphill" edge of both the top plate and the beam. Then trace the profiles of the top plate and beam onto the rafter, cut out the bird's-mouths at each end, and check the rafter's accuracy by setting it in place at several points along the shed's length. Make any corrections needed in the cuts, and use this rafter as a pattern for the other six. Then make plumb cuts at the ends of each rafter, and nail the rafters into place, as for the basic shed.

Gambrel Roof

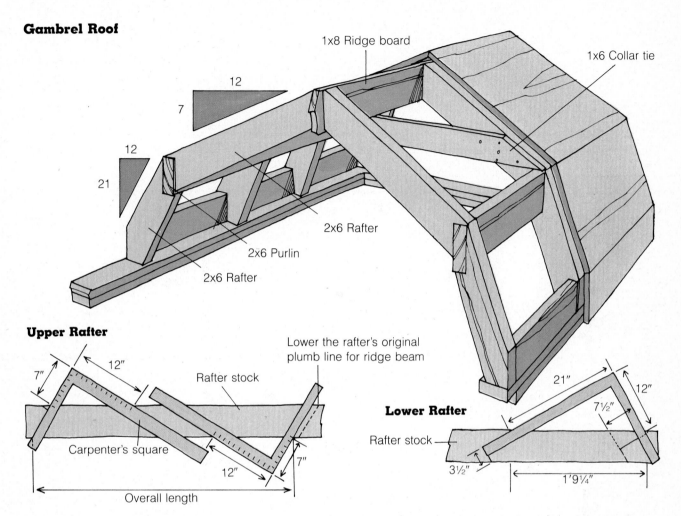

1x8 Ridge board

1x6 Collar tie

12
7
12
21

2x6 Rafter

2x6 Purlin

2x6 Rafter

Upper Rafter

Lower the rafter's original
plumb line for ridge beam

7"
12"
Rafter stock

Carpenter's square

12"
7"

Lower Rafter

21"
12"
7½"
Rafter stock
3½"
1'9¼"

Overall length

The gambrel, or "barn," roof is a common design for sheds. The walls are framed the same way as for the basic shed. Construct the roof by cutting the lower rafters (12 in 21 slope) according to the dimensions in the illustration. Hold the 2 by 8 purlins in place with temporary bracing, and nail the short rafters in place between the purlins and the wall plates. Toenail three 16d nails at the plate connection and face-nail three 16d nails at the purlin. Then cut a pair of pattern rafters for the upper portion of the roof. First, mark the plumb cut at the ridge end by setting a framing square near the end of the 2 by 6 with the 7-inch mark of the tongue and the 12-inch mark of the blade touching the closest edge of the 2 by 6. The blade of the square will then lie across the 2 by 6 at ex-

actly the proper angle for the plumb cut. Mark the 2 by 6 by scribing along the outside edge of the blade. Next, from the ridge cut mark, measure a distance along the edge of the 2 by 6 exactly 3 feet 4 inches, and make a plumb cut mark for the purlin end of the rafter. Mark for the little tang that goes over the top of the purlin by holding a square against the plumb line so the top edge of its tongue intersects the top edge of the rafter exactly 1 1/2 inches away from the plumb line. Scribe along the top edge of the square's tongue, from the plumb line to the edge of the rafter. Finally, before making the actual cuts, go back to the plumb cut line at the ridge end of the rafter and make a new cut line next to it, locating it exactly 1/2 of the ridge board's width to the inside of the original

line. Now cut along this new plumb line, as well as at the purlin end. Then trace and cut out a second rafter and test for accuracy by holding the pair of pattern rafters in place with a scrap of ridge board material between them. Make any necessary adjustments, and mark and cut the remaining six pairs of rafters. Nail each pair in place, starting with the two end ones, to support the ridge board. Toenail three 16d nails at the purlin end, and alternately face-nail or toenail three 8d nails at the ridge board end.

Nail a 1 by 6 collar tie into each pair of rafters with three 8d nails at each end. Then nail 2 by 6 blocking between rafters along the wall plates. Nail 1/2-inch roof sheathing the same way as for the basic shed, and apply roofing.

Alternative Doors and Windows

Hinged Doors

Consider hanging a pair of hinged doors (either a prehung unit or two exterior doors) as an alternative to the sliding door shown with the basic shed. First install the jambs: nail the top and side jambs together, and place the unit into the rough-framed opening. Use shims to square it, then nail the aligned frame in place using 10d finish nails, 2 every 16 inches. Nail the top jamb to the header in the same way. The outside edges of the jambs should be flush with the exterior surface.

To hang the doors, mount three hinges to each door—one 11 inches up from the bottom of the door, one 7 inches from top edge of the door, and one centered between the other two. Mark their locations on the side jambs by temporarily holding the doors in place with enough top clearance so that the doors swing freely. Chisel out recesses to house the hinge faces, and screw the doors in place.

Close the doors until they are flush, mark the inside edge on the jambs, cut the stop material to length, and nail it in place. Cut trim material to length, caulk the back of each piece, and nail the trim casing around the door to conceal the gap between the framing and the jambs.

Cut threshold material to fit between the jambs, and nail or screw it to the floor, overlapping the outside edge of the siding 3/8 inch or so. Then install a cane bolt in one door and either a latch or knob assembly in the other one.

Hinged Doors in Plan

Pre-Hung Doors

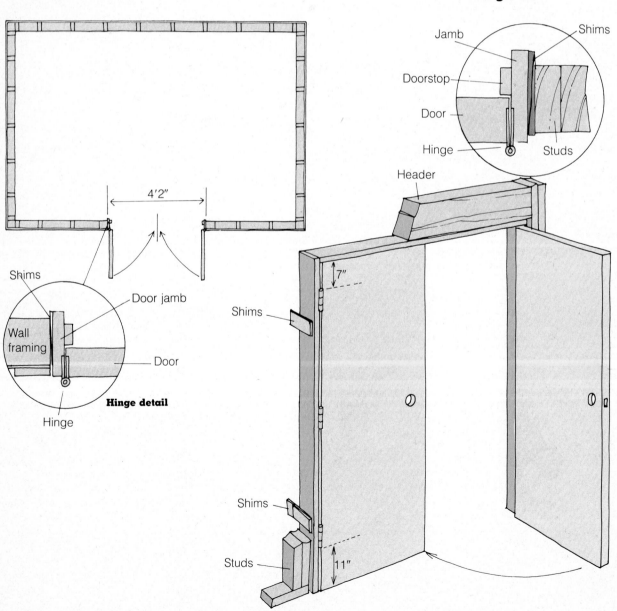

Skylight Windows

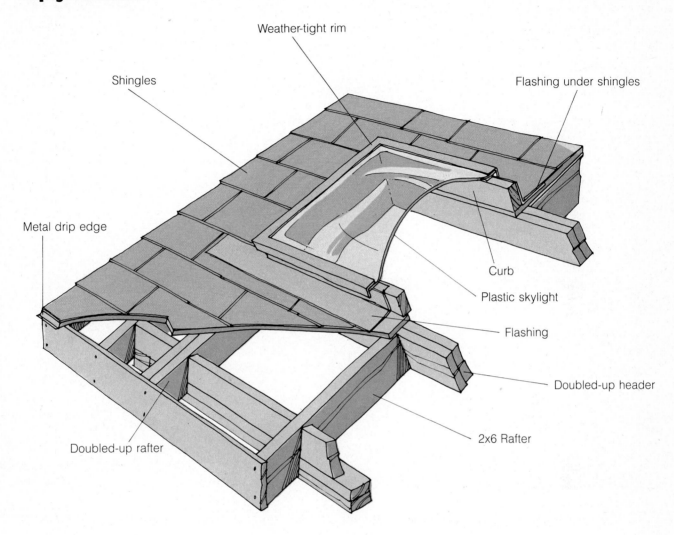

Weather-tight rim

Shingles

Flashing under shingles

Metal drip edge

Curb

Plastic skylight

Flashing

Doubled-up header

2x6 Rafter

Doubled-up rafter

Framing for a skylight is done at the time the rafters are installed. Provide headers at the top and bottom of the skylight by doubling up two pairs of 2 by 6s. Install the outside 2 by 6 of each pair first, nailing each end into an existing rafter with three 16d nails. Attach the tail rafters.

Start by cutting out a section of the rafter that gets "broken" by the skylight. This cutout section should be the same length as the rough opening plus four thicknesses of header material (4 times 1 1/2 inches, or 6 inches). Nail the first 2 by 6 header of each pair, fastening

its ends with three 16d nails face-nailed through each of the full-length rafters. Then face-nail three '6d nails through the single 2 by 6 header into the cut ends of the "broken" rafter. Double up each header with another 2 by 6, and then double up each of the side rafters with another 2 by 6 rafter.

With framing completed, apply sheathing, cutting a hole for the skylight flush with the inside faces of the headers and rafters. Nail a curb made of 2 by 4s around the hole, so it forms a rectangle sitting on top of the roof sheathing. Toenail it with

8d nails every 12 inches. Begin roofing. When you have roofed up to the bottom of the curb, attach collar flashing to the bottom of the curb (the flashing may be included with the skylight unit or can be ordered from a sheet metal shop). Then use step flashing along the sides of the curb as you proceed with the roofing, sliding one piece of L-shaped step flashing underneath each shingle that butts up to the curb. At the top of the skylight, attach the top collar flashing. Finish roofing the shed, and install the skylight unit according to the manufacturer's specifications.

Design Variations

Changing the design elements on a shed can affect how it functions as well as how it looks. The shed's storage capacity and access to things being stored inside are largely set by where you place door and window openings and any interior partitions or walls. These design features also help determine whether the shed is suitable for purposes other than storage and how pleasant or awkward it is to use work areas.

Both the addition of interior walls and changes in the placement of doors and windows are easy changes to make, since the construction steps remain the same. These two pages illustrate two modifications as the examples.

Interior Walls

If your shed is to serve more than one purpose, interior walls can help it do so. Interior walls define spaces clearly and separate activity areas that you don't want to mix. For example, you could run a solid wall through the middle of the shed and place a door to the outside at each end of the structure, to create a playhouse for the kids on one side that is restricted from the tools and chemicals stored on the other side.

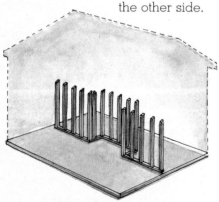

Interior walls can be put up wherever you'd like them. In a space this small, they act as partitions rather than as load-bearing walls. Frame them in the same way as the exterior walls, and nail them to the floor and the wall framing.

Door and Window Placements

Doors and windows can go anywhere, as long as they are properly framed with headers, side jambs, and sills. Modifying a plan to add a window or shift a door isn't difficult. Having them in the right place is important, though, because it will affect the way you enter, move around in, and use the shed.

Choosing the right place is, of course, up to you. The illustrations below and on the facing page show some of the ways in which the location of windows and doors can influence how you use the shed. By experimenting with sketches of your shed plans, you'll be able to see whether a proposed arrangement will let the building function as you want.

Branching Circulation

This shed features a door at the back. A long bank of windows across the front presents an attractive face to the garden. The workbench receives good light and has a pleasant view. The back corners have room for bulk storage, with open and direct access.

Bulk storage

Bulk storage

Access

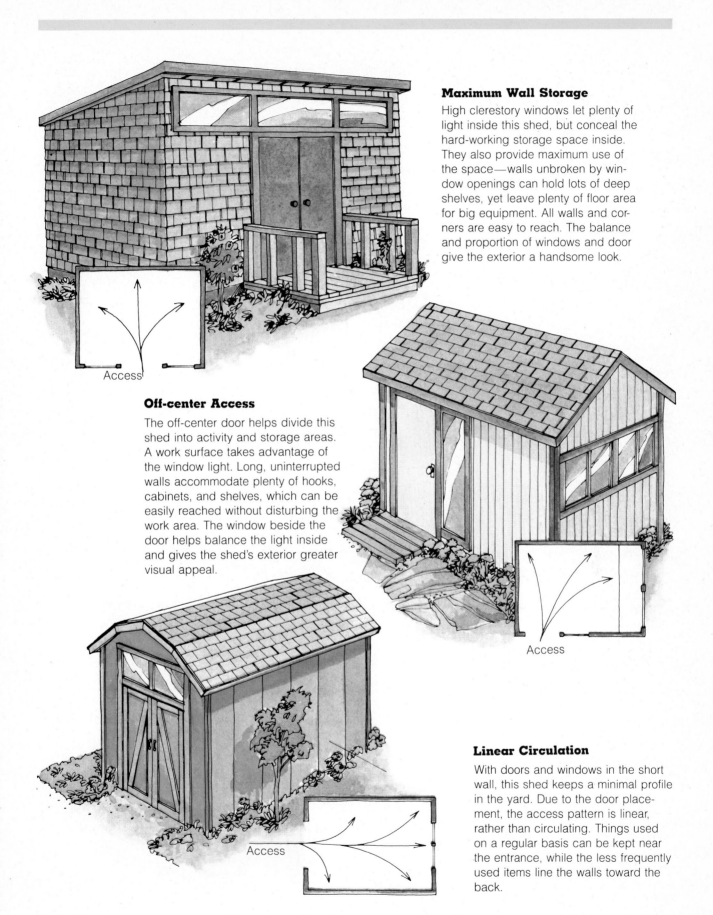

Maximum Wall Storage

High clerestory windows let plenty of light inside this shed, but conceal the hard-working storage space inside. They also provide maximum use of the space—walls unbroken by window openings can hold lots of deep shelves, yet leave plenty of floor area for big equipment. All walls and corners are easy to reach. The balance and proportion of windows and door give the exterior a handsome look.

Access

Off-center Access

The off-center door helps divide this shed into activity and storage areas. A work surface takes advantage of the window light. Long, uninterrupted walls accommodate plenty of hooks, cabinets, and shelves, which can be easily reached without disturbing the work area. The window beside the door helps balance the light inside and gives the shed's exterior greater visual appeal.

Access

Linear Circulation

With doors and windows in the short wall, this shed keeps a minimal profile in the yard. Due to the door placement, the access pattern is linear, rather than circulating. Things used on a regular basis can be kept near the entrance, while the less frequently used items line the walls toward the back.

Access

Three Variations on the Basic Shed

These three sheds are far more than just storage structures. They are also activity centers—efficient work areas or attractive settings for fun and entertaining. Combining functions is a logical way to get the maximum benefit from your shed. Convenience, access, and ease of use are facilitated when storage and the activities it serves share space. By carefully thinking through your needs and developing a personal storage plan, you can come up with solutions, like these, that not only store your possessions neatly but also en-

hance your enjoyment of your home and yard.

These structures all began with the basic shed seen on the preceding pages or a variation of it. Although they serve different purposes, the objectives of each design are the same: a building that handles its intended function well and adds appeal to the landscape rather than detracts from it.

How each shed works is determined by structural details—the placement of a door, the height of a window, the addition of an interior wall. The storage and

work areas are clearly defined, so that the functions don't get muddled.

The attractiveness of these sheds comes from care taken in the choice of finish materials and from the effort made to blend them into their sites. Decks and arbors extend the reach of the structures, keeping them from looking as though they had popped up out of nowhere. These additions also augment living space outdoors, and make your shed both more comfortable and more effective to use as a multi-function activity center.

Pool Cabana

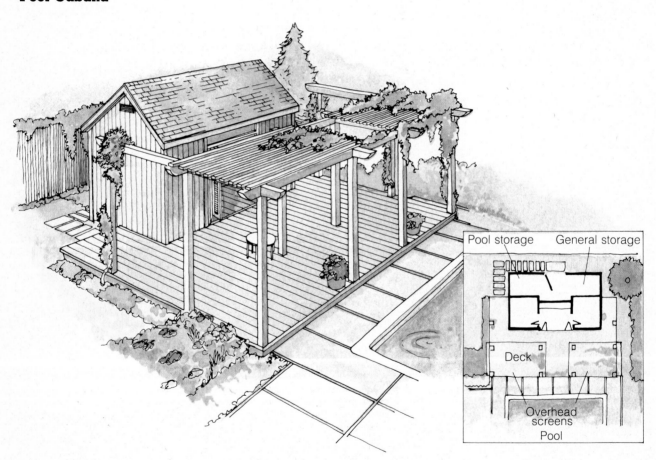

Pool storage | General storage
Deck
Overhead screens
Pool

This poolside structure has been walled through the middle to create two distinct spaces. The front half has two dressing rooms with benches for guests. The high windows give them light while ensuring privacy. Between the dressing rooms, a cabinet holds towels and water toys. The back area has its own entrance. In one half, the me-

chanical equipment for the pool hums away out of sight; the other half houses general household or garden storage. There's also room in the back for chemicals and long-handled pool tools. An arbored patio on the dressing-room side ties the cabana to the pool and provides a comfortable spot to relax out of the sun.

Garden Workspace

This shed is strictly functional—and very pleasant. With its door at the rear, it turns its best face toward the lawn and garden. Wide windows give plenty of light and an agreeable view.

The long workbench offers ample project space, and storage for supplies is close at hand. A drain in the concrete floor makes it easy to water newly potted plants and to hose away dirt.

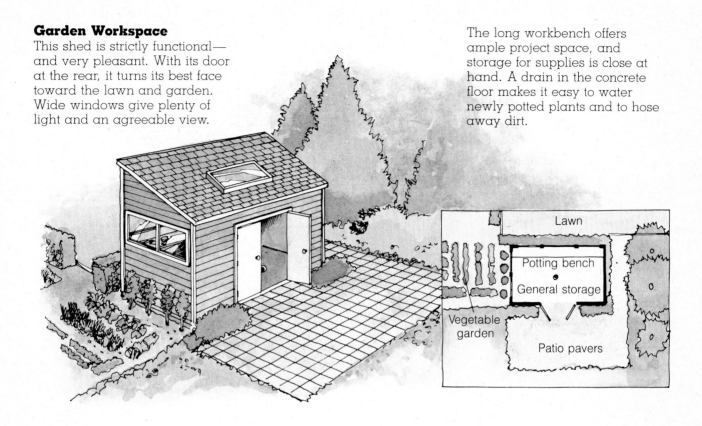

Lawn

Potting bench

General storage

Vegetable garden

Patio pavers

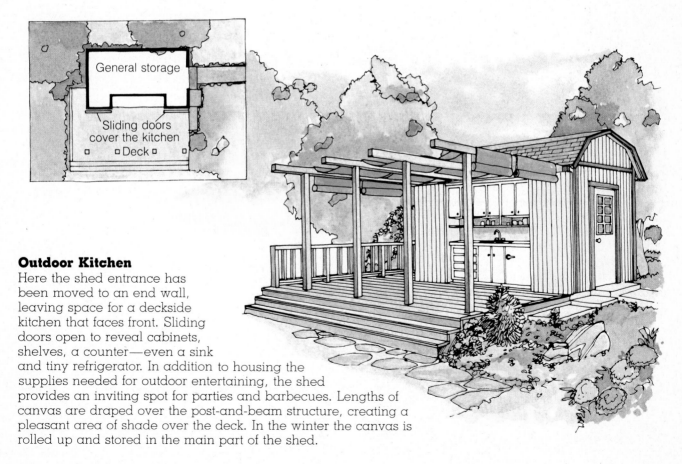

General storage

Sliding doors cover the kitchen

Deck

Outdoor Kitchen

Here the shed entrance has been moved to an end wall, leaving space for a deckside kitchen that faces front. Sliding doors open to reveal cabinets, shelves, a counter—even a sink and tiny refrigerator. In addition to housing the supplies needed for outdoor entertaining, the shed provides an inviting spot for parties and barbecues. Lengths of canvas are draped over the post-and-beam structure, creating a pleasant area of shade over the deck. In the winter the canvas is rolled up and stored in the main part of the shed.

A Lean-To

This small lean-to uses the construction techniques you would use for a larger scale project. You can make it deeper, longer, or higher just by changing the dimensions. Doors can be repositioned, and even windows can be added.

Anatomy of the Lean-To

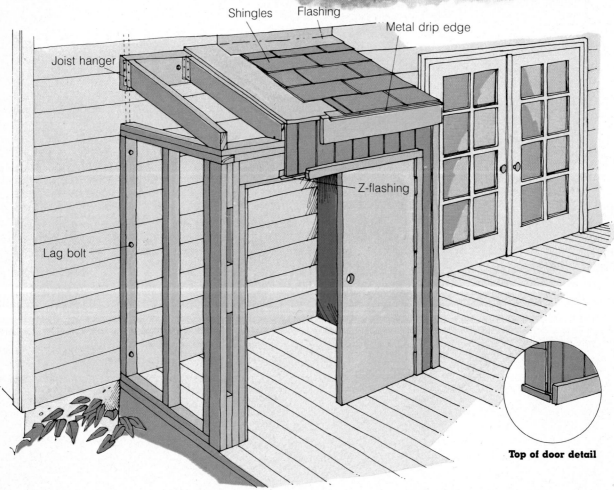

Shingles

Flashing

Metal drip edge

Joist hanger

Z-flashing

Lag bolt

Top of door detail

Wall Construction

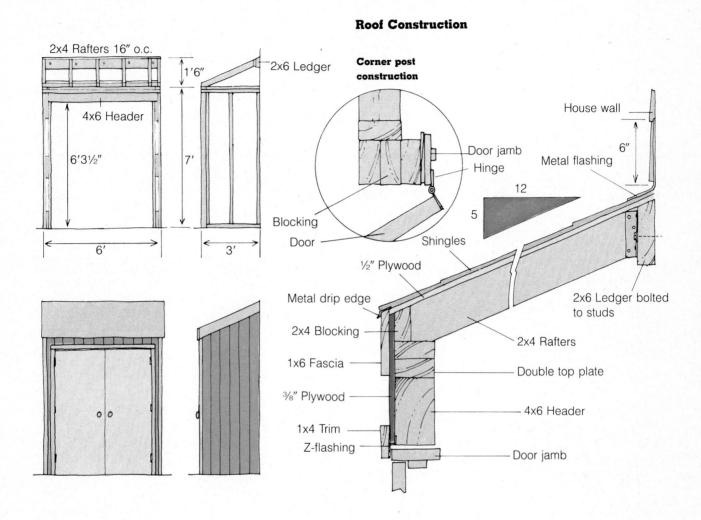

Roof Construction

2x4 Rafters 16" o.c.

1'6"

4x6 Header

6'3½"

7'

6'

3'

2x6 Ledger

Corner post construction

Door jamb

Hinge

Blocking

Door

½" Plywood

Metal drip edge

2x4 Blocking

1x6 Fascia

⅜" Plywood

1x4 Trim

Z-flashing

Shingles

House wall

Metal flashing

6"

12

5

2x6 Ledger bolted to studs

2x4 Rafters

Double top plate

4x6 Header

Door jamb

Construction Steps

1. To lay out the shed, measure and mark 6 feet along the existing building and 3 feet out at both marks. To square the corners, measure both diagonals and adjust the corner positions until the diagonals are equal. Then snap lines on the deck to connect the corners.

2. Build the two end walls, consisting of studs, bottom plate, and one top plate. Raise each wall into place and bolt its end stud into the existing house framing with three 5/16-inch by 5-inch lag bolts. Plumb the walls, and nail the bottom plates into the deck with 16d nails, 16 inches on center.

3. Frame up the front wall with 16d nails and raise it into place. Attach it to the side walls by driving 16d nails through their end studs, 16 inches on center. Nail the side wall cap plates with 16d nails.

4. Cut a 2 by 6 roof ledger to length. Bevel the top edge 22 1/2 degrees. Then bolt it into the house framing with 5/16-inch by 5-inch lag bolts in a staggered pattern. To mark the cuts for a 2 by 4 pattern rafter, hold a short 2 by 4 against one end wall so that its top edge is flush with the top of the ledger at one end, and its bottom edge intersects the inside corner of the front wall's cap plate at the other.

Check it for fit; then cut the rest of the rafters. Attach each one to the ledger with a joist hanger, and toenail it to the cap plate. Then add blocking.

5. Attach roof sheathing, siding, fascia boards, and roofing as for the basic shed. Apply metal flashing where the roof meets the house. Seal with caulking.

6. Hang the doors and attach the trim the same way as for the basic shed. Caulk all exterior joints.

7. If the deck slopes away from the house properly, the shed's floor should stay dry. If not, consider installing a plywood floor in the shed on 2 by 4 sleepers.

A Storage Wall

This raised floor storage structure is built on fence posts. It is intended as a freestanding storage wall that separates distinct areas of the yard and offers access to storage space within from each side. This project is shown with three 4-foot-long sections. It's easy to make the wall longer by continuing to build in 4-foot increments.

Anatomy of the Storage Wall

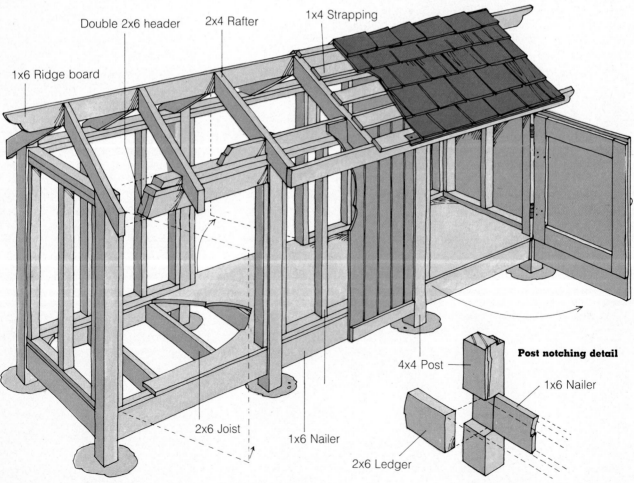

1x6 Ridge board

Double 2x6 header

2x4 Rafter

1x4 Strapping

2x6 Joist

1x6 Nailer

4x4 Post

Post notching detail

1x6 Nailer

2x6 Ledger

Framing Details

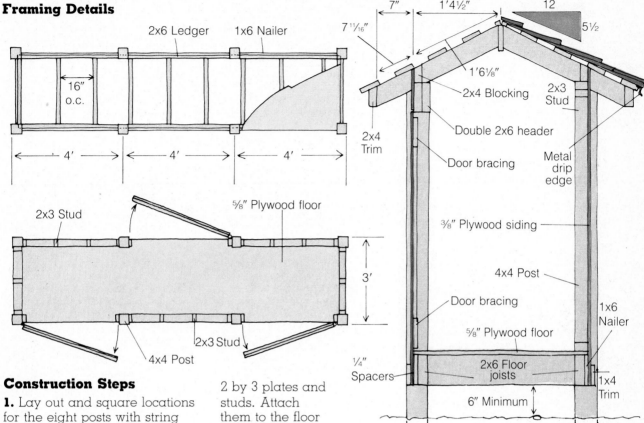

2x6 Ledger 1x6 Nailer

16" o.c.

4'   4'   4'

7" 1'4½" 12

7 ¹¹⁄₁₆"   5½

1'6⅛"

2x4 Blocking   2x3 Stud

2x4 Trim

Double 2x6 header

Door bracing   Metal drip edge

2x3 Stud

⅜" Plywood siding

3'

⅝" Plywood floor

4x4 Post

4x4 Post

2x3 Stud

1x6 Nailer

Door bracing

⅝" Plywood floor

¼" Spacers

2x6 Floor joists

1x4 Trim

6" Minimum

Construction Steps

1. Lay out and square locations for the eight posts with string lines. Dig post holes 24 inches deep and 8 inches across, and place 4 inches of gravel in the bottom. Temporarily brace each post in its hole so that it is plumb and its edge is next to the string. When all eight posts are in perfect alignment, wet the holes, mix concrete, and place it in the holes.

2. Allowing at least 6" clearance to the ground, mark and notch the posts 1 1/2" deep. Nail the two ledgers in place.

3. Cut 2 by 6 floor joists to fit between the ledgers, and face-nail them, 16 inches on center, with three 16d nails at each end.

4. Cut 1 by 6 nailers to fit between the posts on the outside of the 2 by 6 ledgers. Nail them with pairs of 8d nails, 16 inches on center.

5. Nail the 5/8-inch plywood floor, notching it to fit around the posts, with 6d ring-shank nails, 6 inches on center.

6. Frame the solid walls with

2 by 3 plates and studs. Attach them to the floor and posts with 16d nails, 16 inches on center. Add a second cap plate after the walls are in place.

7. To frame the door openings, nail 2 by 3 trimmer studs to both posts, and toenail the header into both the posts and the trimmers with 16d nails, 8 at each end, making the outside flush with the outside of the trimmer studs.

8. Cut 2 by 4 roof rafters, and install them 24 inches on center, with a 1 by 6 ridge board at least 14 feet long. The roof slope is 5 1/2 in 12, or 24 1/2 degrees.

9. Trim the post tops at the same angle as the rafter slope and flush with the rafter tops.

10. Nail 2 by 4 fascia boards to the eave ends of the rafters, with two 16d nails each, leaving at least a 1-foot overhang at each end for decoration.

11. Cut and nail rake rafters to fit between the ridge and fascia boards at the ends.

12. Nail on roof sheathing. For wood shingles, use 1 by 4 boards, 7 inches on center.

13. Install metal drip flashing around the edges, and starting with a double course at each eave, shingle the roof. Use special ridge caps for finishing off the top.

14. Install 3/8-inch plywood siding between the posts. Use 6d galvanized nails, 6 inches on center. Nail a 1 by 4 trim piece across the bottom of each section with 8d galvanized nails, 12 inches on center.

15. Build three doors out of 3/8-inch plywood nailed to 1 by 6 frames bracing the back. Be sure the 1 by 6 frame will clear the inside of the door opening when the door is shut. Hang the doors on heavy-duty butt hinges. Complete the doors by installing concealed latches and a simple pull. Paint or stain the storage wall.

A Carport Enclosure

This type of project can be added under any existing roof—such as a patio overhang or a breezeway. This example uses a carport roof structure to show you how to wall in a new storage space.

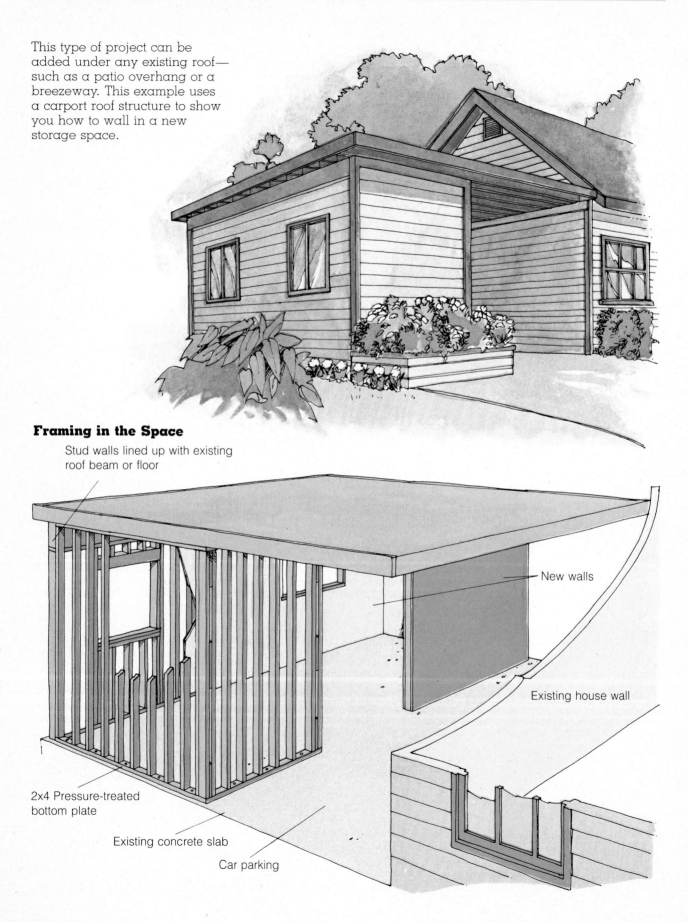

Framing in the Space

Stud walls lined up with existing roof beam or floor

New walls

Existing house wall

2x4 Pressure-treated bottom plate

Existing concrete slab

Car parking

Wall Framing

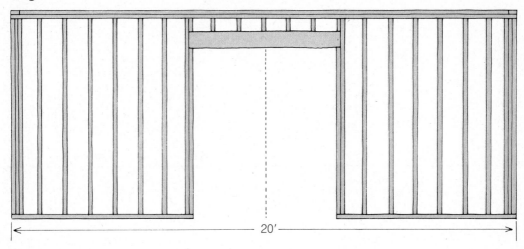

←————————————— 20' —————————————→

Optional Floor Constructions

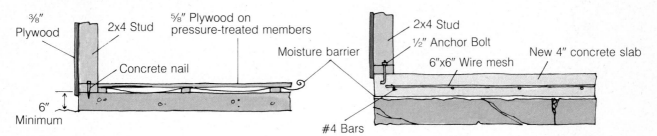

³⁄₈" Plywood

2x4 Stud

Concrete nail

6" Minimum

⅝" Plywood on pressure-treated members

Moisture barrier

#4 Bars

2x4 Stud

½" Anchor Bolt

6"x6" Wire mesh

New 4" concrete slab

Construction Steps

1. Determine whether the existing floor is adequate for storage needs without alteration. It should be free of both surface and subsurface moisture and without extensive cracks. If not, consider pouring a new slab over it or adding a wood floor after the storage walls are built. In most cases, simply coating the existing floor with a waterproof concrete sealer is sufficient preparation. This can be done after the structure is built.

2. Lay out the wall locations on the floor, lining them up with either the edge of the floor or convenient roof framing members. For instance, it may be easier to build a wall directly beneath a roof beam or in line with existing posts, even if it does not line up perfectly with the edge of the floor.

3. Frame the walls, or sections of walls, on the floor with standard stud wall construction. Tilt them up and put them in place, after laying a bead of caulking along the bottom of each plate to seal the joint. Attach the single top plate to an existing roof beam or a cap plate that you have already nailed to the underside of the roof decking. Use 16d nails, 16 inches on center. Attach the bottom plate to the concrete floor with special concrete nails, 2 1/2 inches long. Another method is to shoot the nails into the plate with a powder-actuated nailing gun. This method is dangerous and requires extreme caution. It is also possible to attach the bottom plates by drilling holes for expansion bolts and tightening the bolts down after the walls are in place.

4. Install windows, stapling paper flashing along the bottom and two sides of each window first.

5. Nail on the plywood siding or plywood sheathing for other types of siding.

6. Install the door(s)—either two 3-foot 0-inch hinged doors or a 6-foot 0-inch sliding door fabricated from 2 by 4s and plywood siding.

7. Install screened vents either around the foundation line or near the roof to ensure cross-ventilation.

8. Trim the corners and windows with 1 by 4s or casings to match the existing house trim. Caulk open joints before covering them.

9. Install electrical wiring and fixtures.

10. If a one-hour fire wall is required, nail 5/8-inch type-X wallboard to the interior walls, and tape all the joints.

11. Trim and paint the interior and stain or paint the exterior.

Resources for Storage

Product Manufacturers

Page 26
Storage bar by American
Tack & Hardware

Page 27
Top left: Magnetic holders by
Brookstone Co.
Top center: Hooks by Brookstone Co.
Top right: Tension clips by Graber
Products, Inc.
Center: Magnetic bar by Phelon
Magnagrip Co.; Tension spring by
Hotchkiss Development Co.
Bottom: Bike rack by Graber Products,
Inc.; Tire shelf by E-Z Shelving

Page 28
Left: White plastic crates by Tucker
Housewares; Easy-to-carry boxes by
Corrugated Concepts Ltd.
Right: Red and gray tool boxes by Plano
Molding Co.; Blue tool box by Century
Family Products

Page 29
Open-shelf cabinet by Fort Steuben
Products; Beige cabinet by Sandusky
Metal Products

Page 30
Small red bin by Akro-Mils; Self-stacking
white containers by Ingrid Ltd.; Roll-out
unit by Rubbermaid

Page 31
Top: Both mini-cabinets by Akro-Mils;
White trays by Heller Designs
Bottom: Wall-mounted bin units by
Brookstone Co.; Two-tiered blue holder
by Trophy Products, Inc.; Space
expander drawer by Hirsch Co.; Vinyl-
covered drawer by Rubbermaid

Page 32
Steel mesh metal shelves by Amco; Steel
utility shelves by Hirsch Co.

Page 33
Top: Assorted brackets by Knape & Vogt
Bottom: Red rolling cart by Aikenwood
Corp.; Beige molded plastic unit by
Rubbermaid; White vinyl-covered wire
shelves by Clairson International

Page 34
Storage center by Knape & Vogt

Page 35
Top: Wall-mounted grid by Heller Designs
Bottom: Sliding wire basket by Elfa
West, Inc.

Page 36
Pegboard tool holder by Knape & Vogt

Page 37
Top: Sectioned bins by Trophy Products,
Inc.; Black rack by Merrymaid Plastics
Corp.
Bottom: Red-capped jars by
Brookstone Co.

Addresses of Product Manufacturers

(For further information about products
pictured in this book and about other
storage products, check with your local
home improvement centers or ask them
to write directly to the manufacturers at
the addresses listed below)

Aikenwood Corp.
2151 Park Blvd.
Palo Alto, CA 94306

Akro-Mils
P.O. Box 989
Akron, OH 44309

AMCO
901 N. Kilpatrick Ave.
Chicago, IL 60651

American Tack & Hardware
25 Robert Pitt Drive
Monsey, NY 10952

Brookstone Co.
127 Vosefarm Rd.
Peterborough, NH 03458

Century Family Products
3628 Crenshaw Blvd.
Los Angeles, CA 90016

Clairson International
5100 W. Kennedy Blvd.
Tampa, FL 33609

Corrugated Concepts Ltd.
650 S. Clinton Ave.
Trenton, NJ 08611

Diston Industries
3293 E. Eleventh Ave.
Hialeah, FL 33013

E-Z Shelving
Box 5218
Kansas City, KS 66119

Elfa West, Inc.
14701A Myford Road
Tustin, CA 92680

Gibson Good Tools
P.O. Box 235
Grottoes, VA 24441

Graber Products, Inc.
5253 Verona Rd.
Madison, WI 53711

Fort Steuben Products
Consumer Products Division
10605 Chester Rd.
Cincinnati, OH 45217

Harper-Lee International
308 Prince St.
St. Paul, MN 55101

Heller Designs
41 Madison Ave.
New York, NY 10010

Hirsch Co.
8051 Central Park Ave.
Skokie, IL 60076

Hotchkiss Development Co.
451 Cedar Hill Dr.
San Rafael, CA 94903

Ingrid Ltd.
3601 N. Skokie Hwy.
N. Chicago, IL 60064

Knape & Vogt
2700 Oak Industrial Dr. NE
Grand Rapids, MI 49505

Lobenz-Stevens
460 Park Ave. S
New York, NY 10016

Merrymaid Plastics Corp.
1655 Collamer Ave.
Cleveland, OH 44110

Phelon Magnagrip Co.
East Longmeadow, MA 10128

Plano Molding Co., Plano, IL 60545

Rubbermaid
1147 Akron Rd.
Wooster, OH 44691

Sandusky Metal Products
P.O. Box 1040
Sandusky, OH 44870

Standard Equipment Co.
1900 Emmorton Rd.
Bel Air, MD 21014

Trophy Products, Inc.
9714 Old Katy Rd.
Houston, TX 77055

Tucker Housewares
721 111th St.
Arlington, TX 76011

Other Sources for Project and Shed Plans

American Plywood Association
P.O. Box 11700
Tacoma, WA 98411

California Redwood Association
One Lombard Street
San Francisco, CA 94111

East-Bild Directions Simplified, Inc.
529 North State Rd.
Briarcliff Manor, NY 10510

Hammond Barns
Box 584
New Castle, IN 47362

JKL P.O. Box 33
Fair Lawn, NJ 07410

Louisiana-Pacific Corp.
1300 SW Fifth Ave.
Portland, OR 97201

National Plan Service
435 West Fullerton Ave.
Elmhurst, IL 60126

Osmose Wood Preserving Co. of
America, Inc.
980 Ellicott St.
Buffalo, NY 14209

TECO 5630 Wisconsin Ave.
Chevy Chase, MD 20815

U-Bild Enterprises
P.O. Box 2383
Van Nuys, CA 91409

Western Wood Products Association
1500 Yeon Building
Portland, OR 97204

Other Sources for Shed Kits

Metal Shed Kits

Arrow Group Industries, Inc.
100 Alexander Ave.
Pompton Plains, NJ 07444

Quaker City Industries
301 Mayhill St.
Saddle Brook, NJ 07662

Roper Eastern
9325 Snowden River Parkway
Columbia, MD 21046

Wood Shed Kits

Jer Manufacturing, Inc.
7205 Arthur Dr.
Coopersville, MI 49404

East-West Design Inc.
Box 6022
Madison, WI 53716

United Steel Products
Box 80
Montgomery, MN 56069

Parrott Industries
44 Alco Place
Baltimore, MD 21227

Index

Metric-Conversion Chart

U.S. Measure and Metric Measure Conversion Chart

Formulas for Exact Measures

Rounded Measures for Quick Reference

	Symbol	When you know:	Multiply by:	To find:			
Mass (Weight)	oz	ounces	28.35	grams	1 oz		= 30 g
	lb	pounds	0.45	kilograms	4 oz		= 115 g
	g	grams	0.035	ounces	8 oz		= 225 g
	kg	kilograms	2.2	pounds	16 oz	= 1 lb	= 450 g
					32 oz	= 2 lb	= 900 g
					36 oz	= 2-1/4 lb	= 1000 g (1 kg)
Volume	tsp	teaspoons	5.0	milliliters	1/4 tsp	= 1/24 oz	= 1 ml
	tbsp	tablespoons	15.0	milliliters	1/2 tsp	= 1/12 oz	= 2 ml
	fl oz	fluid ounces	29.57	milliliters	1 tsp	= 1/6 oz	= 5 ml
	c	cups	0.24	liters	1 tbsp	= 1/2 oz	= 15 ml
	pt	pints	0.47	liters	1 c	= 8 oz	= 250 ml
	qt	quarts	0.95	liters	2 c (1 pt)	= 16 oz	= 500 ml
	gal	gallons	3.785	liters	4 c (1 qt)	= 32 oz	= 1 l
	ml	milliliters	0.034	fluid ounces	4 qt (1 gal)	= 128 oz	= 3-3/4 l
Length	in	inches	2.54	centimeters	3/8 in	= 1 cm	
	ft	feet	30.48	centimeters	1 in	= 2.5 cm	
	yd	yards	0.9144	meters	2 in	= 5 cm	
	mi	miles	1.609	kilometers	12 in (1 ft)	= 30 cm	
	km	kilometers	0.621	miles	1 yd	= 90 cm	
	m	meters	1.094	yards	100 ft	= 30 m	
	cm	centimeters	0.39	inches	1 mi	= 1.6 km	
Temperature	°F	Fahrenheit	5/9 (after subtracting 32)	Celsius	32°F	= 0°C	
					68°F	= 20°C	
	°C	Celsius	9/5 (then add 32)	Fahrenheit	212°F	= 100°C	
Area	in²	square inches	6.452	square centimeters	1 in²	= 6.5 cm²	
	ft²	square feet	929.0	square centimeters	1 ft²	= 930 cm²	
	yd²	square yards	8361.0	square centimeters	1 yd²	= 8360 cm²	
	a	acres	0.4047	hectares	1 a	= 4050 m²	